Macromolecular Symposia

Editor-in-chief: Hartwig Höcker
Editors: W. Guth, B. Jung, I. Meisel, S. Spiegel

Symposium Editor: J. Kahovec

139

WILEY-VCH

pp. 1–135

April 1999

Macromolecular Symposia publishes lectures given at international symposia and is issued irregularly, with normally 12 volumes published per year. For each symposium, the Symposium Editor(s) is/are appointed. The journal is produced by photo-offset lithography directly from the authors' typescripts.
Further information for authors can be obtained from
Editorial office "Macromolecular Chemistry and Physics"
Hegelstr. 45/5. OG, D-55122 Mainz, Germany
Tel. +49 61 31/38 17 19; Fax +49 61 31/38 59 28; macromol@wiley-vch.de

More information on our journals, including the "Notes for Contributors to *Macromol. Symp.*", can be found at the following WWW address:

http://www.wiley-vch.de/home/macrosymp

Annual subscription rate 1999
DM 1678.–/SFr. 1408.–/öS 12249.–/US $ 968.– (Europe)/US $ 1108.– (USA/World) incl. postage and handling. Substantial savings are offered by a combined subscription of "Macromolecular Chemistry and Physics" (incl. "Macromolecular Rapid Communications", "Macromolecular Theory and Simulations", "Macromolecular Symposia", "Die Angewandte Makromolekulare Chemie" and "Acta Polymerica"): DM 7768.–/SFr. 6528.–/öS 56706.–/ US $ 4438.– (Europe)/US $ 5128.– (USA/World) incl. postage and handling.

Single issues and back copies are available. Please, inquire for prices.

Orders may be placed through your bookseller or directly at the publishers: WILEY-VCH Verlag GmbH, P.O. Box 10 11 61, D-69451 Weinheim, Tel.: (0 62 01) 6 06–0, Telefax (0 62 01) 60 61 17, Telex 465 516 vchwh d.
E-mail: subservice@wiley-vch.de

Macromolecular Symposia (ISSN 1022-1360) is published with 12 volumes per year by WILEY-VCH Verlag GmbH, P.O. Box 10 11 61, D-69469 Weinheim, Germany. Air freight and mailing in the USA by Publications Expediting Inc., 200 Meacham Ave., Elmont, NY 11003.
Periodicals postage pending at Jamaica, NY 11431.
POSTMASTER: send address changes to Macromolecular Symposia, Publications Expediting Inc., 200 Meacham Ave., Elmont, NY 11003.

Printing: Strauss Offsetdruck, Mörlenbach
Binding: Wilh. Osswald, Neustadt

International Union
of Pure and Applied Chemistry

Macromolecular Division

in conjunction with

Institute of Macromolecular Chemistry
Academy of Sciences of the Czech Republic

Surface and Interfacial Phenomena in Macromolecular Systems

Main and special lectures presented at the
17th Discussion Conference
of the Prague Meetings on Macromolecules
held in Prague, Czech Republic
July 21–24, 1997

Symposium Editor

Jaroslav Kahovec

Organized by the Institute of Macromolecular Chemistry
Director: K. Ulbrich, *P.M.M. Head:* J. Lukáš;
Conference Chairman: A. Sikora

ISBN 3-527-29807-X

The **tables of contents** of the published issues are displayed on the www.

This service as well as further information on our journals can be found at the following www address:

http://www.wiley-vch.de/home/macrosymp

The service is free for everybody.

Contents of Macromol. Symp. 139

IUPAC Symposium on Surface and Interfacial Phenomena in Macromolecular Systems
Prague, Czech Republic 1997

* The asterisk indicates the name of the author to whom inquiries should be addressed.

Author Index

Interfaces Between Coexisting Phases in Polymer Mixtures: What Can We Learn from Monte Carlo Simulations?

Kurt Binder*, Marcus Müller, Friederike Schmid, and Andreas Werner

Institut für Physik, Johannes Gutenberg Universität Mainz, Staudingerweg 7, D-55099 Mainz, Germany

Abstract: Symmetric binary polymer mixtures are studied by Monte Carlo simulation of the bond fluctuation model, considering both interfaces between coexisting bulk phases and interfaces confined in thin films. It is found that the critical behavior of interfacial tension and width is compatible with that of the Ising model, as expected from the universality principle. In the strong segregation limit, only qualitative but not quantitative agreement with the self-consistent field (SCF) theory is found. It is argued that the SCF theory requires $\sqrt{6\chi} << 1$ but for the short chains studied (N = 32 effective monomer units per chain), the limit is only reached for $\sqrt{6\chi}$ close to unity. Also, the effective χ-parameter decreases in the interface. It is shown that the interfacial width w does not increase by the adsorption of block copolymers as long as their areal density is still dilute ("mushroom" regime). But a broadening of interfaces does occur for thin films confined between walls at distance D, due to fluctuations that lead to $w \propto \sqrt{D}$ for short-range forces, in agreement with experiment.

INTRODUCTION

Polymer blends have many useful applications (Refs. 1,2). But since entropy of mixing in a (symmetric) polymer mixture (A,B) with chain lengths $N_A = N_B = N$ is reduced by a factor 1/N (Refs. 3-5), most polymers are not uniformly miscible under conditions of interest. Thus most blends are inhomogeneous on mesoscopic scales, containing A-rich and B-rich domains separated by interfaces (Refs. 1,2,5-7). Structural and thermodynamic properties of these interfaces are both practically relevant and an interesting problem of statistical mechanics (Refs. 5-24). For instance, the following questions arise: how do the interfacial tension σ and the interfacial width w depend on the Flory-Huggins parameter χ and the chain length N? Can one compare theoretical "intrinsic" interfacial profiles with measured ones which are broadened by fluctuations? Does the capillary wave theory account for this

 CCC 1022-1360/99/$ 17.50+.50/0

broadening? Is there an effect of confinement on w in thin films? Is w affected by the adsorption of block copolymers?

We address these problems via Monte Carlo simulation of a coarse-grained model system, the bond fluctuation model (Refs. 18,19,21,23-28) of polymer chains on the simple cubic lattice. Computer simulation of macromolecular materials is a particularly useful tool (Ref. 29) since the model Hamiltonian characterizing intrachain and interchain interactions can be fully specified and all parameters used by phenomenological theories (e.g., Refs. 8-17,20) can be explicitly determined. Thus a stringent test of the theories is possible and, in contrast to experimental tests, no adjustable parameters whatsoever are involved (Ref. 29). Note that systematic errors of simulations (equilibration, finite size effects, etc.) can be controlled for such simplified models and, apart from statistical errors, exact information for the considered model is obtained. Properties are available over a very wide temperature range (by construction, the model shows neither a glass transition at low temperatures, nor chemical degradation at high temperatures); complications, such as asymmetry between A and B, can be eliminated at a first stage (as will be done here) but can be considered at a later stage (Ref. 24). In this way, the physical origin of various complex phenomena can be nicely disentangled. These points will be exemplified in the present paper.

In the next section we introduce the model and summarize its phase behavior in the bulk. Then the temperature dependences of interfacial tension and interfacial width are presented and compared with theoretical predictions (Refs. 8,12,13,17,20). Thereafter, the effect of block copolymers adsorbed at the interface is discussed and finally the effect of confinement on interfacial profiles in thin-film geometry will be analyzed. The paper concludes with a brief outlook on experiments.

MODEL AND BULK PHASE BEHAVIOR

Polymers are represented as effective monomer units connected by effective bonds that can fluctuate in their length (Refs. 25-29), the lengths 2, $\sqrt{5}$, $\sqrt{6}$, 3 and $\sqrt{10}$ lattice spacings being permitted. Henceforth, the lattice spacing is taken as the unit of length. Each effective monomer unit blocks a cube of eight neighboring sites from further occupation, using a simple cubic lattice with a volume fraction of occupied lattice sites $\phi = 0.5$, corresponding to melt densities (Ref. 27). Such a model can be justified (Ref. 29) by integrating 3-5 consecutive C-C-bonds along the backbone of an atomistic chain model into one effective bond. Hence, using a chain length N = 32 (as done here) corresponds to a degree of polymerization of 100-160 in a real polymer. Interactions between monomer units are described by a square well potential of the range $\sqrt{6}$, choosing the most symmetric energy parameters $\varepsilon_{AA} = \varepsilon_{BB} = -\varepsilon_{AB} = -k_B T\varepsilon$.

Apart from the excluded volume, no intrachain potential is used, assuring Gaussian coil statistics (in the melt) even for very small N, i.e., the gyration radius scales as $R_g = b\sqrt{N/6}$, the statistical segment length b being 3.05 ($R_g \approx 7$ for N = 32). The Flory-Huggins parameter χ equals $2z_c\varepsilon$ (Ref. 3), where $Z_c \approx 2.65$ is the effective coordination number in the bulk (Ref. 18). Figure 1 shows the phase diagram of this symmetric model mixture, as obtained from the finite-size scaling analysis (Refs. 18, 28). The critical temperature T_{crit} = 69.35 °C (Ref. 28) is somewhat depressed relative to the Flory-Huggins estimate (χ_c = 2/N, see Ref. 3), $T_c \approx$ 84.40 °C. It can be also seen that the coexistence curve does not have the parabolic shape predicted by the mean-field theory, $\rho_A^{coex} = \frac{1}{2}\left(1 \pm \sqrt{3}\left(1 - \frac{T}{T_{crit}}\right)^{\beta}\right)$ with β =1/2 (Refs. 3-5), but, rather, the critical behavior of the Ising universality implies

$$\rho_A^{coex} = 0.5 \pm 0.682\,(1 - T/T_{crit})^{\beta}, \qquad \beta \approx 0.324. \tag{1}$$

While for $N \to \infty$ the behavior of polymer mixtures is mean-field-like (Refs. 4,5), simulations (Refs. 28,30) and experiments (Ref. 31) have shown that one needs chain lengths at least an order of magnitude longer than those used here to see a crossover towards the mean-field behavior, which hence can be safely disregarded. Of course, a disadvantage of these short chains is that the characteristic length scales of the polymer (b, R_g, and the correlation length ξ of concentration fluctuations) are all of the same order of magnitude (e.g., ξ = 3.5 for ε = 0.03). While simulations with N = 128 are feasible in a strong segregation limit (Ref. 19), in a weak segregation limit this would require a too great computational effort.

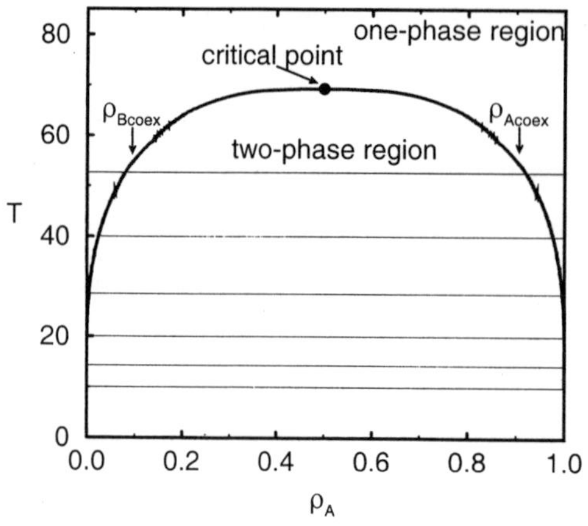

Fig. 1

Phase diagram of the symmetrical A-B polymer mixture (N = 32) with variables temperature (choosing units such as $T = \varepsilon^{-1}$, $k_B = 1$) and volume fraction. The thin horizontal lines show the temperatures where the interfacial properties were investigated (Ref. 18).

INTERFACIAL TENSION AND INTERFACIAL WIDTH

Using finite LxLxL lattices with periodic boundary conditions and applying a multi-canonical (Ref. 32) algorithm in the semi-grandcanonical ensemble, one can sample the minimum $P_L\left(\rho_A^{min}\right)$ of the order parameter distribution $P_L(\rho_A)$ over a wide range of linear dimensions L (Ref. 18). It has been shown (Ref. 33) that for L >> ξ, the system configurations near this minimum are dominated by heterophase states, with two LxL interfaces (connected by the periodic boundary conditions) separating an A-rich domain in the center of the box from A-poor domains surrounding it (or vice versa). Therefore, the depth of the minimum is controlled by the total free energy excess due to these interfaces (Ref. 33), $P_L\left(\rho_A^{min}\right) / P_L\left(\rho_A^{max}\right) = \exp(-2L^2\sigma/T)$. This method takes all interfacial fluctuations fully into account, and does not rely on any direct observation of the interfaces.

A different strategy is needed, of course, for the study of interfacial profiles and the interfacial width w. Here we choose an LxDxL geometry, with D = 64 and L varying from 64 to 512, applying periodic boundary conditions only in the x- and z-directions, while an "antiperiodic" boundary condition in y-direction is used. That is an A(B) monomer that leaves the right boundary of the simulation volume (at y = D/2 = 32) and enters at the left boundary (at y = -D/2 + 1 = -31) as a B(A) monomer; it hence changes its identity. The same holds for movements in the inverse direction. This forces an A-B interface into the system, but does not fix its location. Before taking "measurements", the translational invariance of the system is used to shift the interface to the center, near y = 0. We define $m(y) = 2\rho_A(y) - 1$ from the instantaneous volume fraction profile $\rho_A(y)$ and look for a position y_1 that minimizes the absolute value of the sum $|\Sigma m(y)|$ which runs from $y = y_1 - 20$ to $y = y_1 + 20$ (Ref. 18). Taking "measurements" every 10^4 Monte Carlo steps (MCS) per monomer unit, using the "random hopping" algorithm (Refs. 25-29), with total observation times between $2.5 \cdot 10^5$ MCS and $5.1 \cdot 10^5$ MCS, sufficiently accurate average profiles are obtained. These profiles are fitted by a tanh profile,

$$\rho_A(y) = \rho_0 + \Delta\rho \tanh [(y - y_0)/w], \qquad (2)$$

where $\rho_0 \approx \rho_A^{crit} = ''$, $\Delta\rho \approx \rho_{A,rich}^{coex} - \rho_{A,poor}^{coex}$, as expected from the phase diagram.

In Figs. 2 and 3, the temperature dependences of σ and w obtained in this way are shown. We note that in the strong segregation limit $\left(\rho_{A,rich}^{coex} \to 1, \rho_{A,poor}^{coex} \to 0\right)$, self-consistent field theories give (Refs. 12,13,15,20)

$$\sigma / T = \rho_m b \sqrt{\chi / 6}\,[1 - c(\chi_{crit}/\chi)], \qquad \chi >> \chi_{crit} = 2/N \qquad (3)$$

where ρ_m is the monomer unit density ($\rho_m = 1/16$ here).

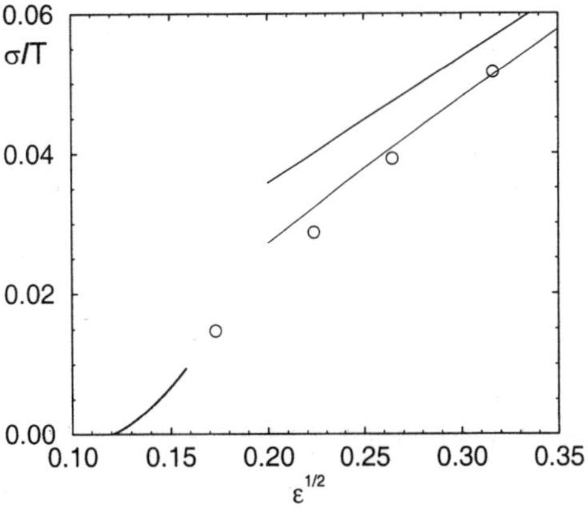

Fig. 2

Interfacial tension σ normalized by temperature T vs. $\varepsilon^{1/2}$ ($\varepsilon = 1/T$). The thick curve near the critical point shows the result of a finite-size scaling analysis: $\sigma/T = 0.0139(\varepsilon/\varepsilon_c - 1)^{2\nu}$, $\nu = 0.63$ being the critical exponent for the correlation length of the 3D Ising universality class. The straight line $\sigma/T = 0.179\varepsilon^{1/2}$ corresponds to Helfand's result in the strong segregation limit ($\chi N \to \infty$) [8]. The thin line shows corrections for the strong segregation behavior by Broseta et al., according to Eq. (3) [13]. (Taken from Ref. 18.)

The constant c of the 1/N correction is still a matter of debate (we use $c = \pi^2/12$ from Ref. 13 here). As we will see below, however, the good agreement between Eq. (3) (using c from Ref. 13) and the simulation data is somewhat accidental: all theories are assumed to be valid only for $\chi_c << \chi << 1$, and this regime is accessible only for very long chains. Thus Fig. 3 shows that w does not agree with the prediction corresponding to Eq. (3), namely (Ref. 13)

$$w = w_{SSL} \{1 + \ln 2\,(\chi_c/\chi)\}, \qquad w_{SSL} = b/\sqrt{6\chi} \quad . \tag{4}$$

Part of this discrepancy is also due to the fact that in the chosen geometry for small w, one does not observe the "intrinsic width" of the interfacial profile considered by the theory (Refs. 8-13), but rather the profile is broadened by fluctuations (see the last section). In the critical region, however, the universal constant $w^2\sigma/T \approx 0.376$ agrees with the Ising model prediction $w^2\sigma/T_c = 0.37$ (Ref. 34) or 0.416 (Ref. 35) rather well, which indicates that one does obtain roughly the intrinsic interfacial width from the simulation in the critical region.

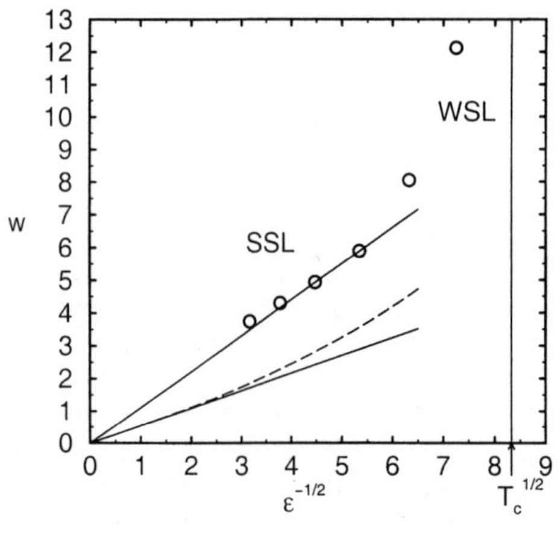

Fig. 3

Interfacial width w vs. $\varepsilon^{-1/2}$ ($\varepsilon = 1/T$). The prediction $w = 0.54\ \varepsilon^{-1/2}$ in the strong segregation limit (Ref. 8) (lower solid line) as well as the prediction of Broseta et al (Ref. 13) (dashed line) are included. The upper straight line shows a fit $w = 1.1\ \varepsilon^{-1/2}$ to the low-temperature data. The vertical line marks T_{crit}. In the weak segregation limit (WSL), the data can be fitted by a critical divergence, $w = \hat{w}\,(1 - \varepsilon_c/\varepsilon)^{-\nu}$ with $\hat{w} = 5.2$, $\nu = 0.63$. (Taken from Ref. 18.)

From Eq. (4), it is easy to understand the above condition ($\chi << 1$) for the validity of the self-consistent field theory: only if $(6\chi)^{-1/2} >> 1$ and w sufficiently exceeds b, the loops formed by the chains across the interface center contain many segments and hence can be treated by a continuum theory. A further effect ignored by the standard treatments is the reduction in the effective coordination number z_c and thus the effective Flory-Huggins parameter χ in the center of the interface (Fig. 4). Thus a reliable theoretical prediction of the intrinsic interfacial width is difficult, even for our simple model where all necessary properties are known.

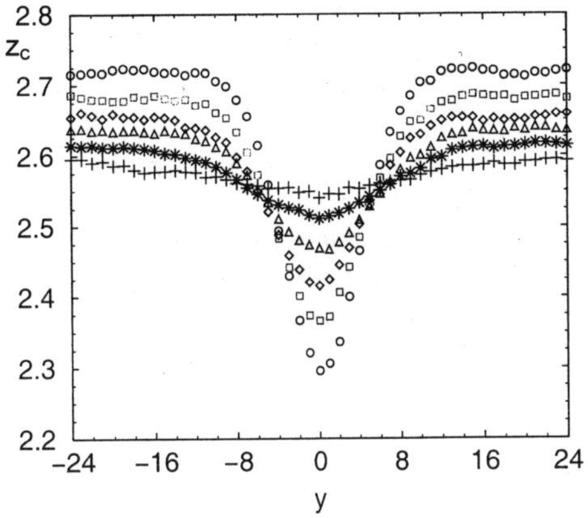

Fig. 4

Effective coordination number z_c plotted vs. the distance y from the center of the interface for N = 32 and several ε. (Taken from Ref. 18.)

o ε=0.1, □ ε=0.07, ◇ ε=0.05, Δ ε= 0.035, ∗ ε=0.025, + ε=0.019

ADSORPTION OF BLOCK COPOLYMERS AT THE INTERFACE

It is rather popular to use block copolymers as polymeric surfactants which adsorb at the interface between coexisting phases of incompatible blends (Refs. 1,2,36), acting as compatibilizers. However, the precise mechanism of this effect is still controversial (Ref. 36). Thus it is of interest to study this problem by our model simulation, too: we can add strictly symmetric and monodisperse block copolymers $A_{fN}B_{(1-f)N}$, f = 0.5, to our A-B interface and vary all parameters of interest in a controlled fashion to study how interfacial properties are affected. Figure 5 shows, as a first step, that in the dilute regime (where the adsorbed block copolymers form "mushrooms" which do not interact with each other), the interfacial profiles of total monomer unit densities are not at all changed, in comparison to a blend at the same value of ε without any copolymers! Since under the conditions of this simulation (1024 chains out of 32 768 were copolymers), about 30% of the interfacial area is covered by block copolymers, this lack of any interfacial broadening by the copolymers is surprising. It turns out, however, that this behavior is in agreement with the corresponding self-consistent field calculations (Ref. 21), which can also account qualitatively, though not quantitatively, for most structural properties of these adsorbed block copolymers. The latter are not yet stretched chains, but rather "dumbbells" consisting of mildly perturbed Gaussian A and B coils, oriented perpendicularly to the interface.

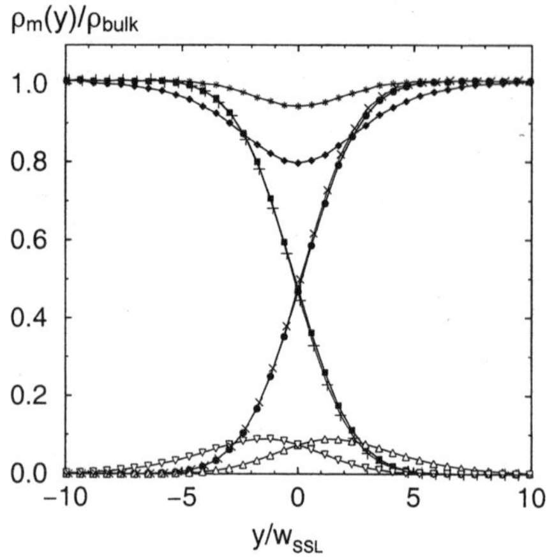

Fig. 5

Monomer unit density normalized by the bulk density, $\rho_m(y)/\rho_{bulk}$ versus y/w_{SSL}. The interfacial width in the strong segregation limit is $w_{SSL} = 1.71$ at $\varepsilon = 0.1$ (cf. Eq.(4)). Densities: total (ρ), ● monomer unit A(ρ_A), ■ monomer unit B(ρ_B), ◆ homopolymers (ρ_h), Δ monomer unit A in the copolymer (ρ_{Ac}), ∇ monomer unit B in the copolymer (ρ_{Bc}), × pure homopolymer A(ρ_A^0), + pure homopolymer B(ρ_B^0). (Taken from Ref. 21.)

SIZE EFFECTS ON INTERFACIAL WIDTHS IN CONFINED FILMS

As a last topic, we consider phase coexistence of an unmixed polymer blend confined between asymmetric walls, a distance D apart, such that the left-hand wall prefers the B-rich phase, while the right-hand wall prefers the A-rich phase. For simplicity, we choose the strength of the wall interactions precisely equal on both walls, $\varepsilon_w = 0.1$, if monomers are at a distance $\Delta z \leq 2$ from either wall (which otherwise is hard and structureless). Choosing $\varepsilon = 0.03$ for $N = 32$, we are far below the critical point ($T/T_{cb} \approx 0.48$), but still above the wetting transition temperature;thus the average interface position is stabilized in the center of the film (Ref. 23). Figure 6 shows that the profile significantly depends on the film thickness. This broadening of the profile with increasing thickness results from capillary waves; the squared width w^2 becomes (e.g., Ref. 17)

$$w^2 = w_0^2 + (4\sigma/T)^{-1} \ln(q_{max}/q_{min}), \tag{5}$$

where w_0 is the "intrinsic width" and q_{min}, q_{max} are the smallest and largest wavenumbers of capillary waves that contribute to the broadening. Taking $q_{max} = 2\pi/\xi_b$, $q_{min} = 2\pi/\xi_{||}$ and estimating (Refs. 22,23) the correlation length $\xi_{||}$ of interfacial fluctuations in this confined geometry as $\xi_{||} \approx \xi_b \exp(\kappa D/4)$, where κ^{-1} is the transverse intrinsic length scale, we find

$$w^2 = w_0^2 + (16\sigma/T)^{-1}\ \kappa D. \tag{6}$$

Taking for w_0 the self-consistent field (SCF) estimate and calculating κ from the theory of Parry et al. (Ref. 37), as quoted in Fig. 6, yields fair agreement with the

Monte Carlo results for large D. For small D, the observed widths fall systematically below the asymptotic estimate, Eq. (6). This happens because then the walls squeeze even the intrinsic interfacial profile.

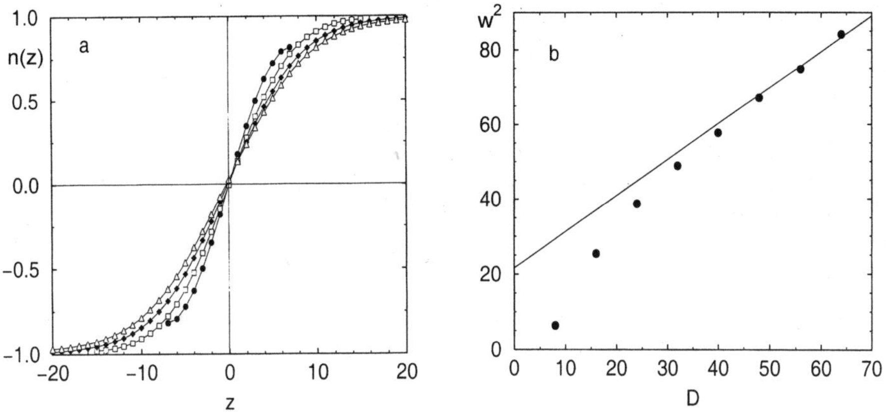

Fig. 6: (a) Order parameter profile $m(z) = (\rho_A(z) - \rho_B(z))/(\rho_A(z) + \rho_B(z))$ versus distance z from the interface center for films of various thicknesses D (lateral size L = 256, periodic boundary conditions in x,y directions). D values: ● 64, ■ 48, ◆ 32, ▲ 16. (b) Squared interfacial width w^2 versus film thickness : • MC simulation, – theoretical prediction (Eq. 16), $\kappa^{-1} = \xi_b(1 + \omega/2)$, $\xi_b = 3.6$, $\omega^{-1} = 4\pi\xi_b^2\sigma/T$, $\sigma/T = 0.015$, $w_0 = w_0^{SCF} = 4.65$). (Taken from Ref. 23.)

CONCLUDING REMARKS

In this paper, it was shown that computer simulations can yield an extremely complete description of interfaces in polymer blends: interfacial tension, interfacial profile of various densities (including particular points along the chains, e.g. chain ends and A-B bonds for block copolymers), and chain configurational properties. In the theoretical description, the SCF theory provides a useful guide in the SSL, while the full crossover to the (Ising-like) critical behavior cannot yet be described theoretically. A factor hampering comparisons of theory with both simulation and experiment is that theory always yields only the "intrinsic" profile, while observed profiles exhibit broadening by capillary-wave fluctuations. The latter are also responsible for a regime in thin films of thickness D, where the width scales as $w \propto \sqrt{D}$ for short-range wall-polymer forces. While this behavior seems to be consistent with some experiments (Ref. 22), other experiments (Ref. 38) claim that $w \propto \ln D$, which is expected for long-range wall-polymer forces (Refs. 23,38). Still others claim (Refs. 39,40) that there is no broadening at all, and one does observe the intrinsic width in experiments, and hence only a preparation-induced "initial roughness" needs to be subtracted. More experiments are desirable to clarify this situation, and we expect that the present work will help to analyze them correctly.

ACKNOWLEDGEMENT

This work was supported in part by the Deutsche Forschungsgemeinschaft (DFG), grant No. Bi314/3, and by the Bundesministerium für Bildung, Wissenschaft, Forschung und Technologie (BMBF), grant No. 03N8008C. We thank J. Klein, T. Kerle, R.A.L. Jones, E.J. Kramer and M. Stamm for stimulating discussions.

REFERENCES

(1) K. Šolc (Ed.), *Polymer Compatibility and Incompatibility - Principles and Practices.* Harwood, Chur 1980.

(2) D.S. Walsh, J.S. Higgins, A. Maconnachie (Eds), *Polymer Blends and Mixtures.* Martinus Nijhoff, Dordrecht 1985.

(3) P.J. Flory, *Principles of Polymer Chemistry.* Cornell University Press, Ithaca 1953.

(4) P.G. de Gennes, *Scaling Concepts in Polymer Physics.* Cornell University Press, Ithaca 1979.

(5) K. Binder, *Adv. Polym. Sci.* **112**, 181 (1994)

(6) I.C. Sanchez (Ed.), *Physics of Polymer Surfaces and Interfaces.* Butterworth-Heinemann, Boston 1992.

(7) K. Binder, *Acta Polym.* **46**, 204 (1995)

(8) E. Helfand, Y. Tagami, *J. Chem. Phys.* **56**, 3592 (1971); **57**, 1812 (1972)

(9) J.F. Joanny, L. Leibler, *J. Phys. (Paris)* **39**, 951 (1978)

(10) J. Noolandi, K.M. Hong, *Macromolecules* **14**, 727 (1981); **15**, 483 (1982)

(11) K. Binder, H.L. Frisch, *Macromolecules* **17**, 2928 (1984)

(12) E. Helfand, S.M. Bhattacharjee, G.H. Fredrickson, *J. Chem. Phys.* **91**, 7200 (1989)

(13) D. Broseta, G.H. Fredrickson, E. Helfand, L. Leibler, *Macromolecules* **23**, 132 (1990)

(14) K.R. Shull, E.J. Kramer, *Macromolecules* **23**, 4769 (1990)
K.R. Shull, Macromolecules **26**, 2346 (1993)

(15) H. Tang, K.F. Freed, *J. Chem. Phys.* **94**, 1572 (1991); **94**, 6307 (1991)

(16) K.R. Shull, A.M. Mayes, T.P. Russell, *Macromolecules* **26**, 3929 (1993)
(17) A.N. Semenov, *Macromolecules* **25**, 4967 (1992); **26**, 6617 (1993); **27**, 2732 (1994)
(18) M. Müller, K. Binder, W. Oed, *J. Chem. Soc., Faraday Trans.* **91**, 2369 (1995)
(19) F. Schmid, M. Müller, *Macromolecules* **28**, 8639 (1995)
(20) A.V. Ermoshkin, A.N. Semenov, *Macromolecules* **29**, 6294 (1996)
(21) A. Werner, F. Schmid, K. Binder, M. Müller, *Macromolecules* **29**, 8241 (1996)
(22) T. Kerle, J. Klein, K. Binder, *Phys. Rev. Lett.* **77**, 1318 (1996)
(23) A. Werner, F. Schmid, M. Müller, K. Binder, *J. Chem. Phys.* **107**, 8175 (1997)
(24) M. Müller, A. Werner, *J. Chem. Phys.* **107**, 10764 (1997)
(25) I. Carmesin, K. Kremer, *Macromolecules* **21**, 2819 (1988)
(26) H.-P. Deutsch, K. Binder, *J. Chem. Phys.* **94**, 2294 (1991)
(27) W. Paul, K. Binder, D.W. Heermann, K. Kremer, *J. Phys. II (Paris)* **1**, 37 (1991)
(28) H.-P. Deutsch, K. Binder, *Macromolecules* **25**, 6214 (1992)
(29) K. Binder (ed.), *Monte Carlo and Molecular Dynamics Simulations in Polymer Science*. Oxford University Press, Oxford 1995.
(30) H.-P. Deutsch, K. Binder, *J. Phys. II (Paris)* **3**, 1049 (1993)
(31) D. Schwahn, G. Meier, K. Mortensen, S. Janssen, *J. Phys. II (Paris)* **4**, 837 (1994)
(32) A. Berg, T. Neuhaus, *Phys. Rev. Lett.* **68**, 9 (1992)
(33) K. Binder, *Phys. Rev.* A**25**, 1699 (1982)
(34) M.E. Fisher, H. Wen, *Phys. Rev. Lett.* **68**, 3654 (1992)
(35) M. Hasenbusch, K. Pinn, *Physica* A (1997, in press)
(36) S.T. Milner, *MRS Bull.* **22**, 38 (1997)
(37) A.O. Parry, C.J. Boulter, *Physica A* **218** 77, 109 (1995)
(38) M. Sferrazza, C. Xiao, R.A.L. Jones, D.G. Bucknall, J. Webster, J. Penfold, *Phys. Rev. Lett.* **78**, 3693 (1997)
(39) M. Stamm, D.W. Schubert, *Annu. Rev. Mater. Sci.* **25**, 325 (1995)
(40) D.W. Schubert, M. Stamm, *Europhys. Lett.* **35**, 419 (1996)

Interfacial Phenomena, Adhesion and Macroscopic Properties in Polymer Composites

M. Pegoraro*, L. Di Landro, F. Severini
Dipartimento di Chimica Industriale e Ingegneria Chimica "G. Natta", Politecnico di Milano, Piazza L. da Vinci 32, 20133 Milano, Italy

Abstract: The microfragmentation or the single-fiber composite method for the evaluation of fiber-matrix adhesion in polymer composites is presented emphasizing advantages and drawbacks. The improvements of the methods introduced in recent years (laser Raman spectroscopy, optical examination of interfacial region, estimation of the stress state at the interface, consideration of viscoelastic properties of the matrix) to obtain more detailed information on the interface failure mechanisms are described. A modified method with a coaxial geometry, which extends the applicability of the test to brittle matrices, is shown. A number of experimental results obtained both with thermoplastic and thermoset matrices containing carbon and glass fibers is reported. Chemical modifications of carbon fiber surfaces were carried out, which effectively changed the adhesion to the high-performance bismaleimide matrix. In particular, the treatment with ammonia is shown to be promising for improvement of adhesion in graphite/bismaleimide composites.

INTRODUCTION

Evaluation of adhesion between polymer matrices and reinforcing inclusions is of remarkable importance in the study of the mechanical behaviour of composite materials. The inclusion/matrix interface is a critical area in determining the composite properties. The importance of the interface is strongly dependent on the chemical nature of the components, on the loading conditions and on the inclusion geometry. Therefore different techniques should be employed for inclusions with the aspect ratio of unity (spheres) and higher (fibers).
Several test methods were developed for measuring adhesion using single fibers or groups of fibers. In the last years three basic methods have been mainly used: fiber pull-out, fiber fragmentation, and fiber microcompression (indentation) [1]. The single-fiber fragmentation is particularly interesting because the state of stress produced on the single fiber is more similar to the state of stress present in a composite than in the other two methods mentioned. Moreover, in transparent polymer matrices it is possible to determine, in transmitted polarized light, the mode and the locus of failure.
We used this method in evaluating the fiber-matrix interfacial adhesion, particularly in polymers with good high-temperature performance, both thermosets and thermoplastics, such as bismaleimide (BMI), epoxy resin, polyetheretherketone (PEEK), polyetherimide (PEI-ULTEM), polyethersulfone (PES), poly(ethylene terephthalate) (PET), polypropylene (PP) [2-6]. We will discuss the main fundamental aspects and the new improvements which make the single-fiber composite (SFC) test one of the most powerful practical methods for the adhesion studies.

Microfragmentation or single-fiber composite test

It is known that a reinforcing effect can be obtained only if the volume fraction of fibers is higher than a minimum value (V_{min}) [7]. Below this value, the fibers break during a tensile test at small elongations without a failure of the whole composite. Let us consider a single

 CCC 1022-1360/99/$ 17.50+.50/0

fiber of diameter d, immersed in a matrix, keeping its volume fraction lower than V_{min}. Under axial load the fiber equilibrium equation gives

$$\tau^*(\pi d)\mathrm{d}x = \left(\frac{\pi d^2}{4}\right)\mathrm{d}\sigma_\mathrm{f} \tag{1}$$

and

$$\tau^* = \frac{d}{4}\frac{\mathrm{d}\sigma_\mathrm{f}}{\mathrm{d}x} \tag{2}$$

where σ_f is the fiber stress and τ is the local shear stress at the fiber surface over a length dx. A theoretical elastic model of the stress σ along a fiber immersed in a polymer and τ at its interface, imposed by an external axial load, was proposed by Cox [8].

As the external load is increased, τ may reach the yielding shear stress of the matrix (τ_y) beginning at the fiber ends in the case of very good adhesion, or reach the maximum interface adhesion shear stress (τ_{ad}) if the latter is lower than the matrix yield shear stress (Fig. 1a). Another failure mode is the frictional debonding.

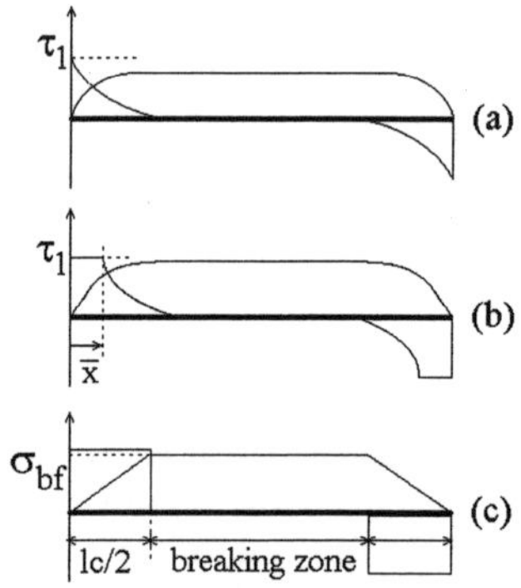

Fig. 1 Stress distribution along a fiber fragment embedded in a matrix

Whatever the mode, a τ limit (τ_l) is assumed to be reached. On increasing strain, a further increase in σ_f is possible because a larger interfacial area at constant τ is involved by the shear phenomena (Fig. 1b) until the fiber strength value σ_{bf} is reached. Fiber failure occurs at or beyond a length defined as semicritical length $l_c/2$ (Fig. 1c). For the fiber length lower than l_c, the tensile strength of the fiber cannot be reached. By integration of Eq. (2) for constant τ, we get the interfacial shear strength (IFSS) expression for τ_{ad}

$$\tau_\mathrm{ad} = \frac{\sigma_\mathrm{bf} d}{2 l_\mathrm{c}} \tag{3}$$

The fiber fragmentation will continue until each fragment is reduced to a length $l \leq l_c$. In conclusion, the fiber will break into fragments of length between $l_c/2$ (minimum) and l_c (maximum), i.e. the average fragment length l will be

$$\bar{l} = \frac{3 l_\mathrm{c}}{4} \tag{4}$$

The lengths of the fragments are measured by the nonius of an optical microscope.

Figure 2 shows, for example, the fragment length distribution $P(l)$ of carbon fibers embedded in polyetherimide (PEI) and epoxy matrix. The mean fragment length $\bar{l}$ is calculated in correspondence of $P = 0.5$; l_c and $_{ad}$ are calculated using Eqs. 3 and 4.

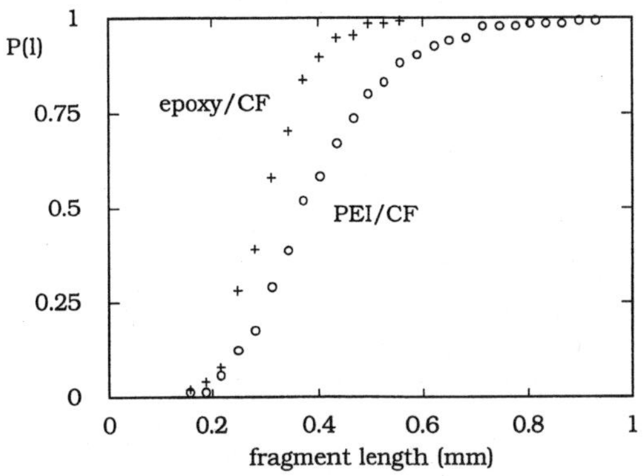

Fig. 2
Examples of fragment length distributions in SFC tests

Table 1 reports a series of results obtained by us in a number of fiber/matrix systems; τ_{ad} values were calculated using σ_{bf} values obtained as in the following.
The actual strength of brittle fibers is distributed on their length over a relatively wide range, the experimental distribution of the fragment lengths being wider than in the case when σ_{bf} is constant along the fiber. l_c depends both on the mean value of σ_{bf} and on its distribution.

Table 1 Adhesion parameters of matrix/fiber systems

Matrix/fiber system	l_c (mm)	τ (MPa)
PEEK/carbon (HTA, XAS)	0.52	37.3-39
PEI/ carbon (HTA, XAS)	0.55-0.5	35-40
PES/carbon (HTA)	0.76	23.5
PP/glass	3.02	3.75
Epoxy (Shell Epon828)/carbon (AS4)	0.5	47.9
Epoxy (Shell Epon828)/graphite (HMS4)	0.52	40.3
Epoxy (DGEBA-IPDA/carbon (XAS)	0.63	36
Epoxy (Ciba LY556-HT972)/carbon (T800)	0.41	45.7
Epoxy (Ciba LY556-HT972)/graphite (M40)	0.41	31
Epoxy (Ciba LY556-HT972)-BMI/T800	0.43	38.5
Epoxy (Ciba LY556-HT972)-BMI/M40	0.70	17.5
Toughened epoxy (DGEBA-IPDA+2.5% CTBN)/XAS	4.5	5
Epoxy (DGEBA-IPDA)/glass	0.5	56
Toughened epoxy (DGEBA-IPDA+2.5% CTBN)/glass	1.3	12

HTA - Toho Rayon, XAS - Courtalds, T800, M40 - Toray, AS4, HMS4 - Hercules

The tensile strength of brittle fibers such as carbon and glass fibers, behaves according to the Weibull distribution function [9]. The failure probability $P(\sigma)$ of a fiber for stresses lower than σ is expressed as

$$P(\sigma) = 1 - \exp\left[-l\left(\frac{\sigma}{\beta}\right)^{\alpha}\right] \tag{5}$$

and the mean fiber strength $\bar{\sigma}$ of a large number of fibers is [5]

$$\overline{\sigma} = \frac{\beta}{l^{1/\alpha}} \Gamma\left(\frac{\alpha + 1}{\alpha}\right) \tag{6}$$

where l is the fiber length, α and β are the Weibull parameters and Γ is the complete gamma function.

A number of adhesion studies was done by many authors, following the microfragmentation procedure [10]. It was found that the proposed method gives rather reproducible results, which are useful for comparative and practical uses. Its employment is especially important when surface modifications or different sizing agents have to be studied.

The SFC method presents a number of theoretical and practical difficulties. In fact it always assumes an elastoplastic behaviour of the matrix and/or of the interface and the presence of a constant τ in correspondence of the fiber breaks even when failure happens by a mixed failure mode. Moreover, it is not able to put in evidence the thermodynamically reversible work of adhesion because of the irreversible mechanical dissipations. The Kelly-Tyson model is not able to distinguish friction effects during debonding from the elastoplastic shear in the bonded zone. During a fragmentation test, in the debonded zone, τ depends on pressure only ($\tau = \mu p$ where μ is the friction coefficient and p is the radial stress at the fiber surface). In the actual situation, the shear stress is not equal for all the interface failure mechanisms; therefore the maximum adhesion stress τ_{max} is greater than the mean stress τ estimated according to the Kelly-Tyson procedure. Another practical limitation is that the test can be applied to matrices characterized by ultimate strains equal at least 2÷3 times the fiber break strain ε_{bf}. This is needed to reach saturation in the fragmentation process.

The τ_{ad} appears to be a phenomenological parameter rather than a well founded absolute measure of the adhesion. However, many efforts are currently done to try to put in evidence separately the different physical and chemical factors which influence τ_{ad} and to express quantitatively these contributions.

The last main improvements are:

(i) the direct method of fiber strain measurement;

(ii) the examination of the failure types in the SFC tests and the study of the corresponding energies needed for failure;

(iii) the study of the viscoelastic behaviour of the matrix which rules the pressure at the interface and the debonding.

In the following we present some considerations on these new aspects of the SFC test.

Direct method of fiber strain measurement

A very important method for measuring the interfacial shear strength (IFSS) is nowadays the laser Raman spectroscopy (LRS) [11]. It is capable of measuring the strain ε of the fiber immersed in a polymer matrix in each point of the fiber, when submitted to an external load. In this way, the stress σ present in the fiber can be evaluated by $\sigma = E\varepsilon$. The elastic modulus E must be obviously found with the fiber alone by preliminary tensile tests.

Carbon fibers (but not glass fibers) are suitable for strain analysis by Raman laser spectroscopy. They exhibit three distinct bands in the Raman spectrum. In the first-order region, up to 1800 cm^{-1}, scattering effects are found at about 1350 cm^{-1} (D line) attributed to vibrations induced by disorder and at 1580 cm^{-1} (G line) attributed to E_{2g} in-plane vibration of the graphitic cell; in the second-order region, the band around 2700 cm^{-1} (G' line) is associated with the three-dimensional crystallinity and is related to the 1350 cm^{-1} vibration. Low-modulus high-tenacity fibers (such as T800) are less crystalline than high-modulus fibers (such as M40) and present broader Raman D and G bands not exhibiting the G' line. The G line is normally used for strain analysis, being particularly reliable for crystalline high-modulus fibers. Raman frequency of the G peak was found to shift in a regular proportional way towards lower frequency values by increasing tensile strain. A linear correspondence between Raman frequency shift (cm^{-1}) and strain is found by calibration [12]. The spatial resolution of LRS is of the order of 1 μm. Confocal Raman microscopy allows to obtain the spectrum of the carbon fiber embedded in a resin.

When the single-fiber composite sample is submitted to a tensile experiment, microfragmentation is developed and each broken segment end becomes a place of stress variation at the ends from 0 to the σ_f. The elementary force balance equation Eq. (1) allows to express the local shear stress acting on the fiber in the place x at the interface (ISS) as

$$\tau(x) = E_f \frac{r_0}{2} \frac{d\varepsilon(x)}{dx} \tag{7}$$

Therefore the Raman method allows to measure for the first time the axial stress and the shear stress distribution in the proximity of each segment end directly and not only by using mathematical models (such as that of Kelly and Cox).

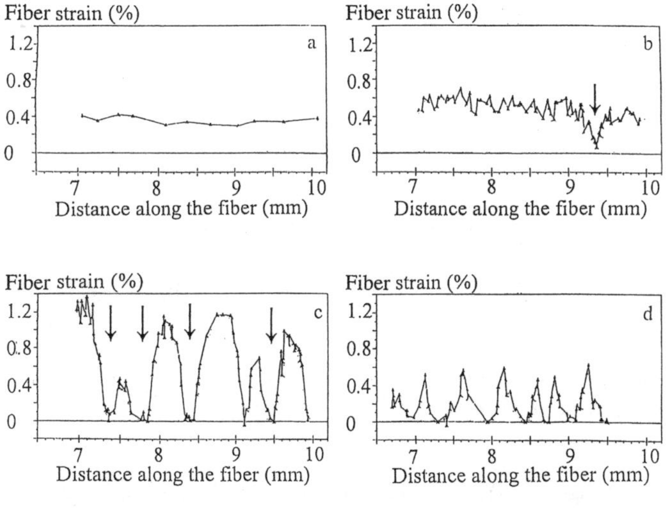

Fig. 3 Fiber strain measured by laser Raman spectroscopy during a microfragmentation test (applied strain: a 0.6, b 0.8, c 1.5, d 3.0 %)

Figure 3, obtained by Galiotis [13] by operating on the system Toray M40 sized fiber, Narmco 5250-2 BMI, and Ciba LY556-HT972 epoxy shows the fiber strain as a function of the distance of the examined points from the origin of the chosen fragmentation gauge length. At low mean strain, the local strains are uniformly distributed over x (Fig. 3a); when the applied mean strain increases, the first fracture is induced (Fig. 3b) but the local damage does not influence significantly the strains of the other points of the fiber and does not affect the stress increase borne by the rest of the fiber due to the increased elongation. Going on in increasing the mean strain, other fractures arise until saturation is reached; the strains in different fragments remain high in their central parts but are strongly reduced in the places where new fractures are formed (Fig. 3c). After saturation going to higher applied strains, all

the fiber fragment strains are strongly reduced (Fig. 3d): this is mainly attributed to the interface damage (debridging).
By examining the $\varepsilon(x)$ profile around a fracture, it is possible to calculate the corresponding ISS (τ) distribution along x. The maximum value of the derivative of the function $\varepsilon(x)$ corresponds to the maximum of τ. The maximum ISS at low $\bar{\varepsilon}$ is normally at the fragment ends (Fig. 4), which agrees with the Cox equation. Beyond a certain applied strain (over 1.5 %), however, the maximum of ISS appears away from the fiber fragment end and shifts towards the middle of the fragment on increasing the applied strain. All the ISS maxima named IFSS data are averaged on all the fragments at each applied strain and their mean value is calculated as a function of strain (Fig. 5). The average maximum of ISS goes down when, increasing the applied elongation, interfacial debridging takes place. The maximum value of IFSS before debonding results which is somewhat above that calculated by the Kelly-Tyson microfragmentation method at saturation (which is a mean value).

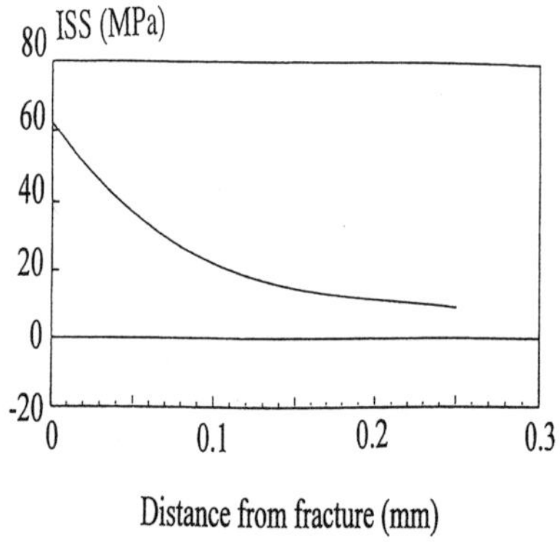

Fig. 4 Shear stress distribution at the fiber – matrix interface (applied strain 1.2 %)

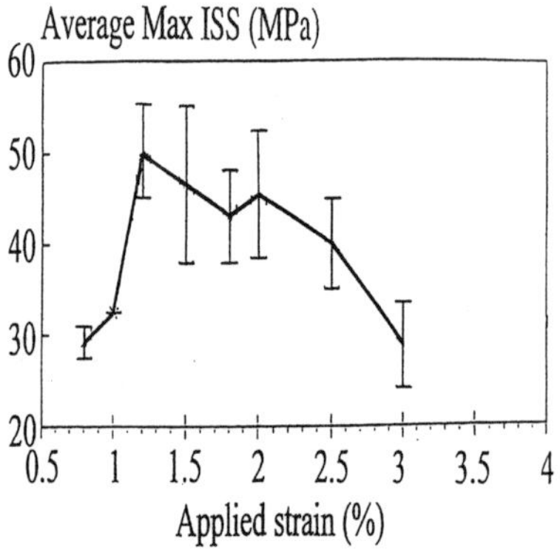

Fig. 5 IFSS as function of applied strain during SFC tests

Obviously, operating by LRS, the Kelly-Tyson calculation of τ_{ad} can be obtained by measuring the average fragment length, $\bar{l}$, by calculating l_c and by using for σ_{bf} the proper value obtained by the Weibull equation. The comparison gives IFSS = 46.6±8.7 MPa and τ_{ad} = 38 MPa [13].
The LRS method is able to evidence the debonded length of each fragment where ISS is zero. Figure 6 shows that these lengths increase with the applied strain, initiate at 1.5 % strain and involve 5 % of the fiber length.

Fracture shape and energetic aspects of SFC test

The shape of the fracture observed in polarized light in transparent systems during and after the tensile tests can clarify the failure mechanism at least in a qualitative way. Figure 7 shows examples of [14]:

(a) well developed stress field around the fiber breaks, which demonstrates an efficient shear mechanism of load transfer;

(b) fiber break and cone- or disk-shaped failure in the matrix, close to the fiber ends;

(c) easy debonding.

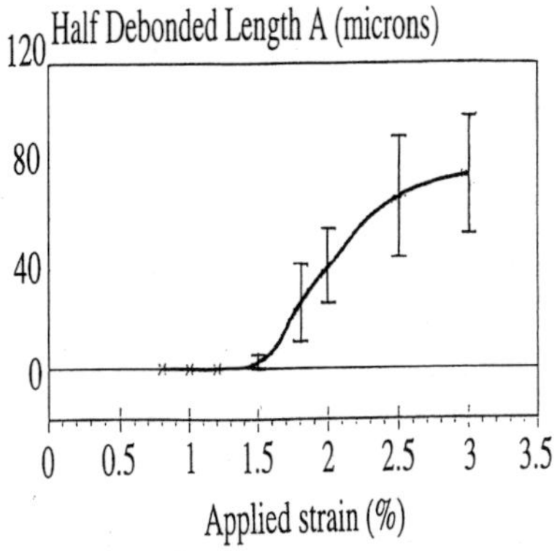

Fig. 6 Debonding length as function of applied strain during SFC tests

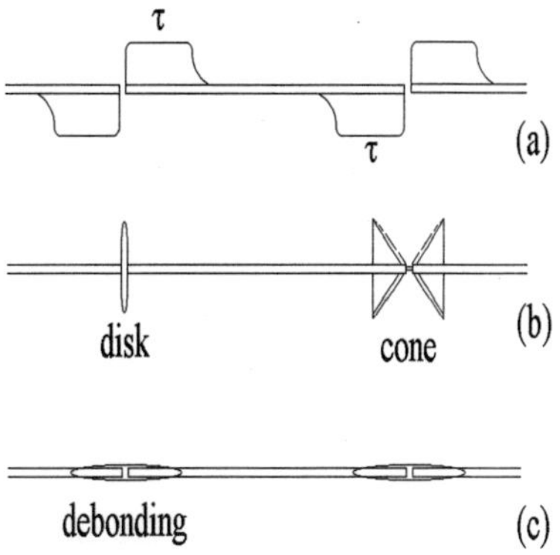

Fig. 7 Schematic of possible interface failure mechanisms

Table 2 shows the relation between the bond type, strength and the mechanism of fracture in cases (b) and (c).

Table 2 Relation between modes of failure and failure mechanisms

Type of failure	Properties
Failure starting even at low strains from filament ends in a strong matrix develops: - disk-shaped crack - two inclined conical cracks - both or none, shear strains evident	High bond strength
Failure starting from filament ends in a weak or soft matrix develops: - inclined conical cracks in the matrix	Bond strength sufficient to load filament by transfer
Shear stress pattern shifts toward filament center during loading. Friction stresses may act in the debonded region.	Bond strength not sufficient to load the filament to fracture.

To better understand the results obtained by the SFC test, one can take into consideration the energy involved in fracture in different components and in the region of the interface. Experimental observations show that when failure occurs, considering an instantaneous constant strain ε, the total elastic energy of the system ($U_m + U_f$) is at least in part spent in different possible ways, which are mainly:

(1) fiber cracks formation ΔU_{fb};

(2) formation of cracks in the matrix (disk-shaped or conical surfaces starting generally from both ends of the broken fiber) (ΔU_{mb});

(3) debonding energy associated with an interface or interphase rupture running around the fiber surface (ΔU_{deb});

(4) energy release in vibrations and viscoelastic dissipations (ΔU_d).

The mechanical energy balance when no external work is applied is expressed by

$$-\Delta\left(U_f + U_m\right) = \Delta U_{deb} + \Delta U_{fb} + \Delta U_{mb} + \Delta U_d \tag{8}$$

where U_f and U_m are the elastic energies of fiber and matrix and the other ΔU are defined above in an explicit way. Some of these terms can be expressed rather easily.

(1) We try now to express ΔU_f taking into consideration a fragment length l_0 of the fiber under stress σ_{bf} (Fig. 8).

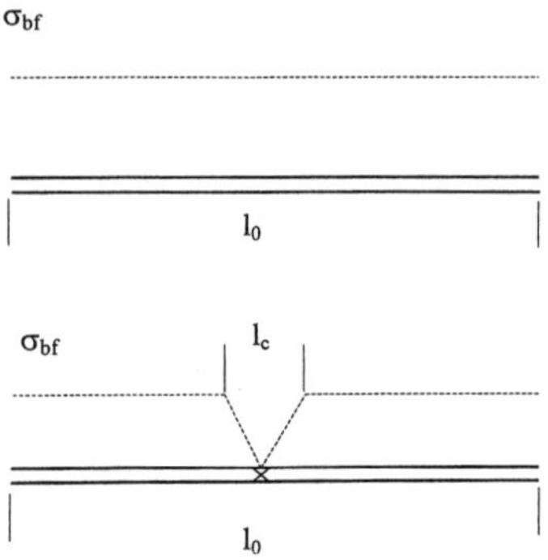

Fig. 8 Variation of stress distribution in a broken fiber

After the break, two fragments are generated. The system is now characterized by segments of length $l_c/2$ and segments of length $(l_0 - l_c)$ which remain in the same stress conditions as before. The elastic energy of the fiber segment (length l_0) before break is

$$U_f^0 = \frac{1}{2}\sigma_{bf}\varepsilon_{bf}\left(\frac{\pi d_f^2}{4} l_0\right) = \frac{1}{2}\frac{\sigma_{bf}^2}{E_f}\frac{\pi d_f^2}{4} l_0 \tag{9}$$

The final energy of the undisturbed zone is

$$U_f' = \frac{1}{2}\frac{\sigma_{bf}^2}{E_f}\frac{\pi d_f^2}{4}\left(l_0 - l_c\right) \tag{10}$$

For our considerations the energy state of reference is zero when $\sigma = 0$. The energy present in the two newly broken fiber ends is

$$U_f'' = \frac{1}{3}\frac{\sigma_{bf}^2}{E_f}\left(\frac{l_c}{2}\right)\frac{\pi d_f^2}{4} \tag{11}$$

In fact the expression for U_f'' (elastic energy in the critical length of the fragments) related to one fiber fracture is obtained using the hypothesis that τ is a constant in the range $0 \le x \le l_c/2$ so that the behaviour of stress and strain is linear with respect to x. The equations are:

$$\sigma(x) = 2\frac{\sigma_{bf}}{l_c}x \tag{12}$$

$$\varepsilon(x) = \frac{\sigma(x)}{E_f} = \frac{2\sigma_{bf}}{E_f l_c}x \tag{13}$$

The energy $dU_f''/2$ accumulated in each fragment end at a distance x from the origin in the volume element $dV = A\,dx$ is

$$\frac{dU_f''}{2} = \frac{1}{2}\frac{\sigma^2(x)}{E_f}dV = \frac{\pi d_f^2}{2E_f}\frac{\sigma_{bf}^2}{(l_c)^2}x^2\,dx \tag{14}$$

Integrating (for the two fragments),

$$U_{\mathrm{f}}'' = -\frac{\pi d^2}{E_{\mathrm{f}}} \frac{\sigma_{\mathrm{bf}}^2}{(l_{\mathrm{c}})^2} \frac{\left[x^3\right]_0^{(l_{\mathrm{c}}/2)}}{3} = \frac{1}{3} \frac{\pi d_{\mathrm{f}}^2}{4} \frac{l_{\mathrm{c}}}{2} \frac{\sigma_{\mathrm{bf}}^2}{E_{\mathrm{f}}} \tag{15}$$

Finally,

$$-\Delta U_{\mathrm{f}} = U_{\mathrm{f}}^0 - \left(U_{\mathrm{f}}' + U_{\mathrm{f}}''\right) = \frac{l_{\mathrm{c}}}{3} \frac{\pi d_{\mathrm{f}}^2}{4} \frac{\sigma_{\mathrm{bf}}^2}{E_{\mathrm{f}}} \tag{16}$$

i.e.,

$$-\frac{\Delta U_{\mathrm{f}}}{U_{\mathrm{f}}^0} = \frac{2}{3} \frac{l_{\mathrm{c}}}{l_0} \tag{17}$$

This fiber energy decrement associated with fracture according to Eq. (8) is balanced by the energy spent by the different ΔU_i whichever is possible.

For example at low mean strains, if matrix and interface are strong, an increment of ΔU_{m} is mainly expected, as shown by the strong effects in polarized light due to shear stresses which are evidenced around the fragment ends. At higher strains, we observed that debonding may appear; the stress σ in the fiber observed by LRS becomes lower and interfacial shear stress is around zero demonstrating a failure of the interface. A complete study of energy partition can be done at low strains by using the finite element elastic model of the SFC test [14].

A complete observation of all the sample points is required to look at all the failure steps; simultaneous recording of stress, strain, and work is also needed to discover the values of different ΔU_i and the related stresses.

(2) Energy spent for the disk-shaped surface in the matrix (diameter D). Starting from the linear elastic fracture mechanic theory, we put $K_{\mathrm{IC}} = \sqrt{E\ G_{\mathrm{IC}}}$ where K_{IC} is the critical first-mode stress intensity factor at which fracture is propagated ($K = \sigma\sqrt{\pi a}$; a is the fracture initial length equivalent to the fiber diameter when fracture starts). Therefore

$$\frac{\pi D^2}{4} \frac{K_{\mathrm{IC}}^2}{E} = \Delta U_{\mathrm{b}} \tag{18}$$

In the case of a cone formation, $\Delta U_{\mathrm{bm}} = GA$ where A is the cone surface and G is in principle not the same G_{IC} because the fracture mode is not simply the first.

(3) The debonding energy can be expressed as

$$\Delta U_{\mathrm{deb}} = \overline{G}_0 \pi d_{\mathrm{f}} l_{\mathrm{deb}} \tag{19}$$

where $\overline{G}_0$ is the mean specific energy needed to detach the unit interphase surface and l_{deb} is the debonding length.

If τ is taken as a constant during the debonding (when we find a yield process at the interface), the force needed to detach the fiber is

$$F = \tau \pi d_{\mathrm{f}} l_{\mathrm{deb}} \tag{20}$$

Considering the energy variation due to the debonding advancement from x to x+dx, we get

$$\mathrm{d}U_{\mathrm{deb}} = \frac{F}{l_{\mathrm{deb}}} x \,\mathrm{d}x \tag{21}$$

By integrating we get

$$\Delta U_{\mathrm{deb}} = \tau \pi d_{\mathrm{f}} \int_{0}^{l_{\mathrm{deb}}} x \,\mathrm{d}x \tag{22}$$

i.e.,

$$\Delta U_{\mathrm{deb}} = \tau \pi d_{\mathrm{f}} \frac{l_{\mathrm{deb}}^2}{2} \tag{23}$$

Therefore the mean specific energy is

$$\overline{G}_0 = \frac{\Delta U_{\mathrm{deb}}}{\pi d_{\mathrm{f}} l_{\mathrm{deb}}} = \frac{\tau l_{\mathrm{deb}}}{2} \tag{24}$$

Effects of viscoelastic matrix on adhesion factor τ_{ad}

The pressure existing at the interface is one of the main factors which rule adhesion at different temperatures both in physical and chemical interphase bridging. Outwater suggested a simple relation valid in the debonding case, $\tau = \mu\sigma_{\mathrm{R}}$, where μ is a friction coefficient. The radial stress σ_{R} can be calculated from different models on the basis of the composite geometry. For a single continuous cylindrical inclusion, as in the SFC test before breaks, the Nairn model [15] can be used; a simplified expression gives [16]

$$\sigma_{\mathrm{R}} = \frac{(\alpha_{\mathrm{m}} - \alpha_{\mathrm{f}}) \cdot \Delta T \cdot E_{\mathrm{m}}}{1 + \nu_{\mathrm{m}} + (1 - \nu_{\mathrm{f}}) \dfrac{E_{\mathrm{m}}}{E_{\mathrm{f}}}} \tag{25}$$

which becomes

$$\sigma_{\mathrm{R}} \approx (\alpha_{\mathrm{m}} - \alpha_{\mathrm{f}}) \cdot \Delta T \cdot \frac{E_{\mathrm{m}}}{1 + \nu_{\mathrm{m}}} \tag{26}$$

in the case $E_{\mathrm{f}} \gg E_{\mathrm{m}}$.

These expressions show that pressure depends on the difference between the mean dilatation coefficient of matrix (α_{m}) and of fiber (α_{f}), on the temperature variation ΔT after the molding or curing, and on the elastic properties (E_{m}, E_{f}, ν_{m}, ν_{f}) of the components.

In the case of polymer matrices, properties E_{m} and ν_{m} are functions of temperature and time. Therefore calculations of stresses must take into consideration their viscoelastic behaviour. This is done first of all by constructing master curves $E(t)$ and $\nu(t)$ or $G(t)$ at a reference temperature T_0 and the shift factors $a_{T_0}^{T}$. The master curve gives the proper modulus at T_0 as a function of a reference time ζ defined as

$$\zeta = \frac{t}{a_{T_0}^{T}} \tag{27}$$

With the master curve, the relaxation modulus can be evaluated in any condition if the cooling history $T(t)$ is known. Pressure can be therefore calculated according to Nairn or by Eq. (25).

Procedure of stress calculation at the interface

Let us start with cooling from a reference temperature T_r at which thermal stresses are considered to be zero. For a thermoplastic, one may assume $T_r = T_g$ for amorphous polymers and $T_r = T_m$ for crystalline polymers. For thermoset, the cure temperature may be taken as T_r. It is convenient to split the total temperature difference T_r-T into small equal steps ΔT_j (j =1, 2, ... u). In the SFC, for each ΔT_j, we get a radial contraction Δr_j which generates the corresponding $\Delta\sigma_j$ relaxing according to the matrix viscoelastic behaviour. E_m and ν_m values are found on the master curves as functions of the reference time ζ. A simple example illustrating the calculation procedure is shown in Fig. 9 (three cooling steps).

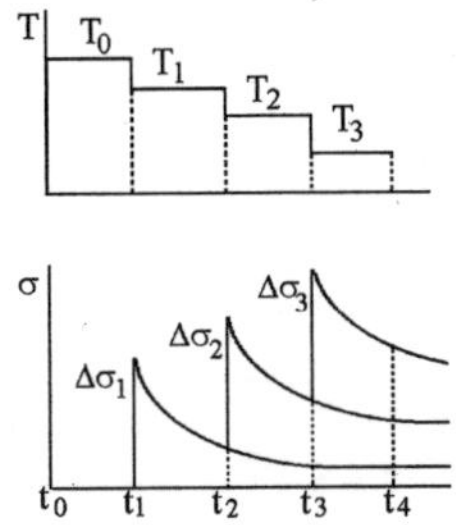

Fig. 9 Schematic of the cooling process

At time t_4 are present the three stresses $\Delta\sigma_1(t_4)$, $\Delta\sigma_2(t_4)$, $\Delta\sigma_3(t_4)$ generated, respectively, at times t_1, t_2, t_3 and relaxed in the time intervals t_4-t_1, t_4-t_2, t_4-t_3. The calculation program which solves the Nairn model equations is used to calculate the radial stress $\Delta\sigma_3$ (t_4) (and also the hoop and axial stresses).

This is done using the E_m and ν_m values obtained from the master curves at a reference time $\zeta_3 = \frac{(t_4 - t_3)}{a_{T_3}}$ and the expansion coefficients $\alpha_m(T_3)$. The next step is to calculate $\Delta\sigma_2$ (t_4), which was generated at time t_2, relaxed at T_2 during the reference time $\zeta_2 = \frac{(t_3 - t_2)}{a_{T_2}}$ and subsequently relaxed at T_3 during the reference time ζ_3. The Nairn equations are solved again using E_m and ν_m obtained from the master curves at the reference time $\zeta_2+\zeta_3$ and the proper α_m. Analogously it is possible to calculate $\Delta\sigma_1(t_4)$ making reference to E_m and ν_m valid for $\zeta_1+\zeta_2+\zeta_3$ where $\zeta_1 = \frac{(t_2 - t_1)}{a_{T_1}}$.

The total stress at the final time of observation (t_u) on the basis of the superposition principle is in general

$$\sigma(t_u) = \sum_j \Delta\sigma_j(t_u - t_j) \tag{28}$$

The example of Fig. 10 shows σ_R at the interface of the system CF-PEI resin of a SFC sample, calculated using the Nairn model assuming an elastic behaviour of the fiber and a viscoelastic behaviour of the PEI matrix (for cooling rates of 10 and 1 °C/min). For comparison, the behaviour of the elastic matrix and the elastic fiber is also shown.

Different master curves of thermosets such as bismaleimide (BMI) and epoxy matrices and of thermoplastic polymers, both amorphous such as PEI and crystalline such as PEEK were prepared by our group and were used for these calculations [3,17]. The calculations give an important information on the radial stress entity but experimental check of the σ_R is not yet feasible at present. However, the importance of thermal stress relaxation on adhesion can be shown, for example, by the critical lengths measured at room temperature for the epoxy DGEBA-IPDA and PEI with high-tenacity carbon fibers (XAS), cooled down at different cooling rates [3]. It appears clearly (Fig. 11) that high cooling rates bring shorter l_c and higher τ_{ad}; this agrees with our relaxation model which shows that higher interface pressures and

τ_{ad} (Eq. (25) are reached at higher cooling rates (Fig. 10).

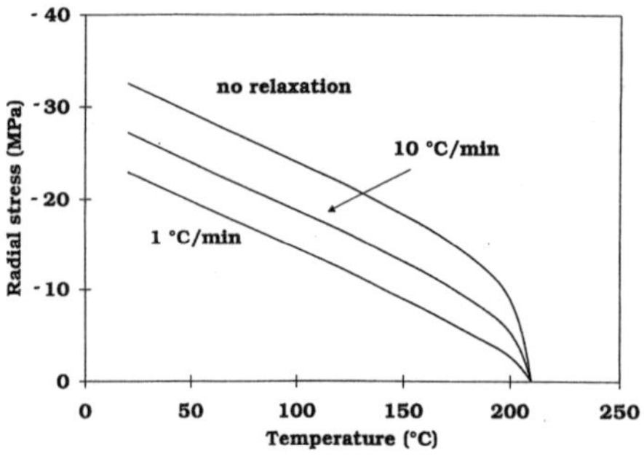

Fig. 10 Radial residual stress at the PEI/XAS interface as function of the end cooling temperature

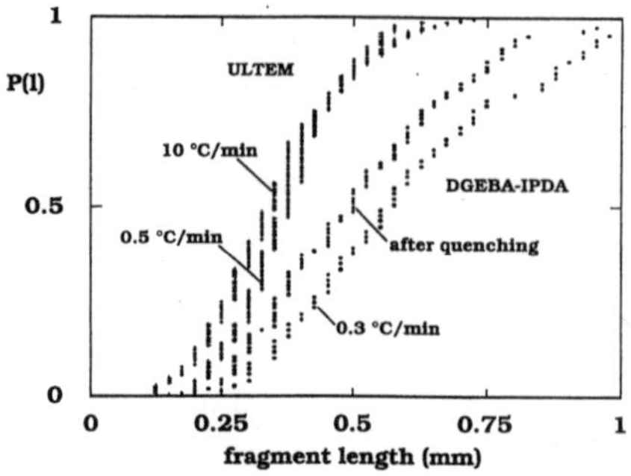

Fig. 11 Fragment lengths observed in epoxy/XAS and PEI/XAS SFC tests after coling at different rates

Equation (25) indicates that pressure at the interface must decrease by increasing the temperature for two reasons: the first is that factor ΔT is reduced, the second is that the modulus decreases on increasing T. SFC tests made at higher than room temperatures in fact show a regular decrease in τ_{ad} both for thermoplastic and thermoset resins. Figure 12 shows the behaviour of the thermoplastic ULTEM (amorphous) and PEEK (semicrystalline) in contact with the same carbon fiber (Courtalds XAS). It can be seen that in thermoplastic matrices, adhesion goes remarkably down with increasing temperatures and reaches a negligible value for temperatures approaching the T_g (amorphous polyetherimide ULTEM) and T_m in the case of PEEK (T_g = 143 °C, T_m = 334 °C). Epoxy matrices (CIBA LY556 + HT972 and DGEBA-IPDA) with different carbon fibers (Toray M40 and Courtalds XAS) show a strong decrease of adhesion on increasing T but, in correspondence with the cure temperature (140 °C) and even above, a residual adhesion is maintained.

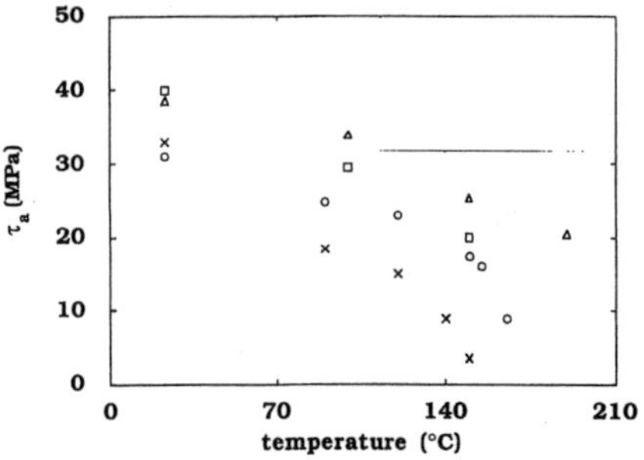

Fig. 12 τ_{ad} as function of temperature in different composite systems (□ ULTEM-XAS, Δ PEEK-XASO, O LY556/HT972-M40, X DGEBA/IPDA-XAS)

The presence of a consistent value of $\tau = \tau_c$ at 140 °C is an indication of the existence of a chemically bonded structure which should be independent of temperature. The difference $\tau_{ad} - \tau_c$ therefore should be attributed to physical effects and mainly to the pressure generated by the different thermal strains of matrix and fiber and by the Poisson effect. The last one gives a contribution to pressure given [16] by

$$\sigma_{\mathrm{RP}} = \frac{\varepsilon(\nu_{\mathrm{m}} - \nu_{\mathrm{f}})}{\left[\frac{(1+\nu_{\mathrm{m}})}{E_{\mathrm{m}}} + \frac{(1-\nu_{\mathrm{fT}} - 2\nu_{\mathrm{fL}}^2)}{E_{\mathrm{fT}}}\right]} \quad (29)$$

where ν is the Poisson coefficient; subscripts m and f refer to matrix and fiber; L and T indicate longitudinal and transversal directions.

An estimate for an epoxy-CF system shows that σ_{RP} at 25 °C is about 30 % of the total σ_R [18].

Extension of the SFC method to brittle matrices

When the matrix is so brittle that the strain-at-break of the matrix is lower than the fiber break strain, $\varepsilon_{bm} < \varepsilon_{bf}$, the fiber-fracturing process generating fragments cannot take place and no critical length can be measured. A case of our recent interest was that of BMI (Narmco 5250-2, BASF) (ε_r = 1.7%) used with high-tenacity carbon fibers, unsized TUS (ε_r = 1.9%). The general mechanical properties of BMI and fibers are reported in Table 3.

Table 3 Mechanical properties of matrices and fibers

Quantity (unit)	BMI	Epoxy (LY556-HT972)	TUS	MUS	XAS
σ_r (MPa)	69	65	~ 5000	~ 3000	4500
E (GPa)	3.8	2.4	260	330	230
ε_r (%)	~ 1.7	> 5 %	~ 1.9	~ 0.9	~ 1.9

A fragmentation method applied to a three-phase monofilament named coaxial SFC, useful to get shear adhesion stress τ_{ad}, was first suggested by Jacques and Favre [19]. It is mainly useful for comparing the efficiency of different adhesives with the same adherent. They proposed to coat the fiber with a thin sheath of the brittle resin (BMI) and to embed the coated fiber into a ductile resin (epoxy) which maintains the specimen integrity during the tensile test (Fig. 13).

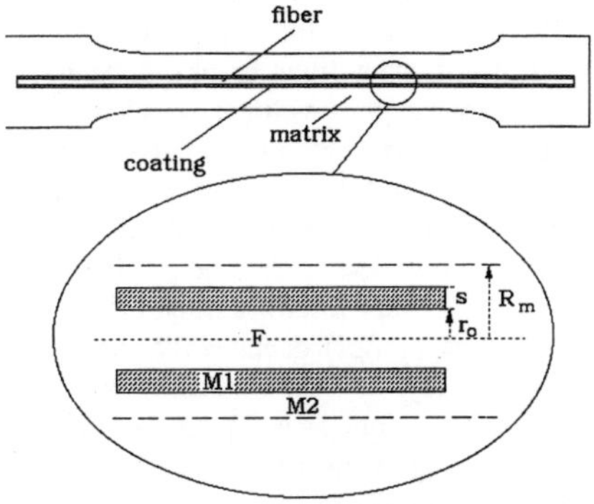

Fig. 13 Geometry of the coaxial single fiber composite coupon

When the brittle resin sheath M1 is broken, the load becomes shared between the fiber F bound to M1 and the tough resin M2 until the fiber is broken somewhere on increasing the load. At this moment, the load exchange between the tough matrix and the fiber end is obtained, as usual, by the shear stress transfer which is transmitted through the tough and the brittle resins to the fiber. In this way, fragmentation can take place but the failure mechanical study is quite difficult. We suppose that shear yielding or debonding should happen more probably at the interface F/M1.

Stress, strain and shear energy analysis of a system made by a continuous fiber embedded in a matrix with or without an intermediate sheath can be done according to the mechanics of continuous elastic systems [15, 17]. With a computer program, different failure criteria in any point of the system can be checked by a relative comparison of the local values with the possible limit values of materials and interfaces. In our case, the highest values of von Mises stress, of the strain energy density W_E and of the distortion energy density W_D, are found at the interface between fiber and BMI (Fig. 14). The stress calculation was done from the data of Table 3 for a load of 50 MPa and a cooling range ΔT of -120 °C.

The introduction of the BMI interlayer considerably reduces radial stresses (σ_R) and tangential stresses (σ_T) in the outer epoxy matrix in comparison with the stresses acting in the same epoxy in the absence of BMI (Fig. 15), when the matrix touches the fiber. We observe that the application of the external load increases the stresses σ_R and σ_T of thermal origin at both the interfaces F/M1 and M1/M2. In conclusion, according to the calculations, the first break in a continuous system should occur in the brittle resin starting at the interface F/M1. After that the load increment on the fiber can soon cause the fiber fracture. Crack propagation in the matrix M1 is stopped at the tough resin interface. It was observed [19], for example in the case of TGDDM (Narmco 5208) and the high-tenacity CF, that fractures in the brittle matrix have a penny-like shape. The load transfer involves shear stresses through the interfaces M2/M1 and M1/F. In this way, the fiber can be re-stressed, new fiber breaks appear and more energy per unit volume is released. In proper conditions, this may also cause the failure at the fiber-BMI interface (debonding). Our observations under an optical microscope show that in our case (TUS fiber/BMI/LY556 + HT972), both microfragmentation and debonding of the fiber occur while the sheath remains steady in the matrix. We can conclude that, in our case, Eq. 3 can be used to calculate τ_{ad} of the fiber as a good means to assess interface properties even if this value is the result of combination of two different processes of shear transfer.

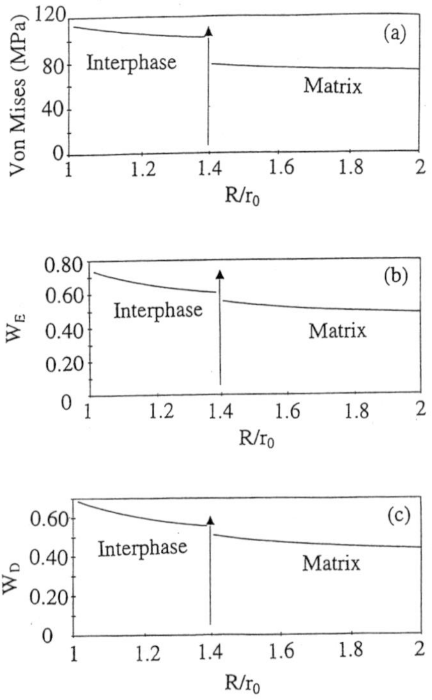

Fig. 14 Comparison of failure criteria in a CF/BMI epoxy system

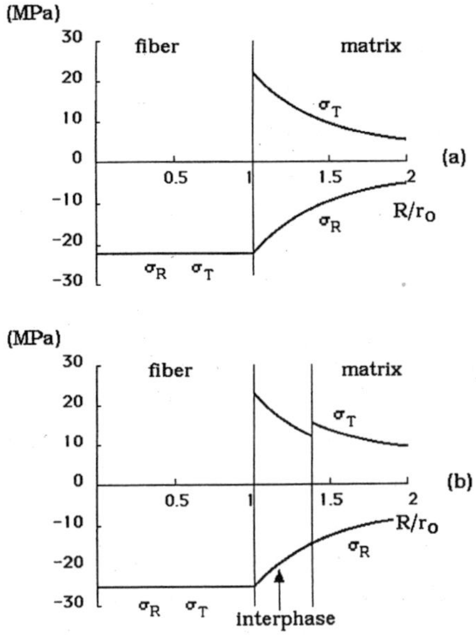

Fig. 15 Stress distribution in a CF/epoxy system with and without a BMI interphase

According to the Cox model [8], the shear stress at the fiber interface embedded in a monophasic matrix is

$$\tau = \frac{E_\mathrm{f}\varepsilon\beta\sqrt{\dfrac{A_\mathrm{f}}{4\pi}}\sinh\beta\left(\dfrac{1}{2}-x\right)}{\cosh\dfrac{\beta l}{2}} \tag{30}$$

where

$$\beta = \sqrt{\frac{G_\mathrm{m}}{E_\mathrm{f}}\frac{2\pi}{A_\mathrm{f}\ln\dfrac{R_\mathrm{m}}{r_0}}} \tag{31}$$

R_0 and R_m are the fiber and the equivalent matrix radii, respectively. When a two-phase matrix is present, we propose as a first approximation and very roughly, to substitute for G_m the equivalent modulus G_c of the series combination of the two matrices. G_c is found by applying the equation

$$\frac{1}{G_\mathrm{c}} = \frac{\phi_1}{G_1} + \frac{\phi_2}{G_2} \tag{32}$$

where ϕ_1 is the volume fraction of the brittle phase M1 and ϕ_2 is the complementary fraction of the tough phase M2.
By simple geometric considerations,

$$\phi_1 = \frac{s(2r_0 + s)}{\left(R_\mathrm{m}^2 - r_0^2\right)} \tag{33}$$

We can therefore calculate from Eqs. (30) and (31) τ_c, which should represent the effective mean adhesion. If the adhesion of fiber with the brittle matrix alone is to be evaluated, Eqs. (30) and (31) should be used employing G_1 modulus for G_m, thus obtaining τ_a. For the ratio $\tau_\mathrm{a}/\tau_\mathrm{c}$, one obtains

$$\frac{\tau_\mathrm{a}}{\tau_\mathrm{c}} = \frac{\beta_\mathrm{a}}{\beta_\mathrm{c}}\frac{\sinh\beta_\mathrm{a}\left(\dfrac{1}{2}-x\right)}{\sinh\beta_\mathrm{c}\left(\dfrac{1}{2}-x\right)}\frac{\cosh\dfrac{\beta_\mathrm{c}l}{2}}{\cosh\dfrac{\beta_\mathrm{a}l}{2}} \approx \frac{\beta_a}{\beta_c} \tag{34}$$

where

$$\frac{\beta_a}{\beta_c} = \sqrt{\frac{G_1}{G_c}} \tag{35}$$

Therefore τ_a should be higher than the τ_c of the composite model.

In conclusion, the experimental SFC τ_a evaluated by the Kelly-Tyson method using a two-phase matrix appears to correspond to the composite model and therefore underestimates the effective adhesion coefficient τ_a between the brittle component alone and fiber.

$$\frac{\tau_a}{\tau_c} = 1.195$$

An example of the application of the modified SFC method in our recent studies of adhesion to the system with high-tenacity and high-modulus carbon fibers (T800 and M40) and BMI Narmco 5250-2 embedded in epoxy Ciba LY556+HT972 is shown. In our case G_1 = 1.62 GPa, G_2 = 1.04 GPa, G_c = 1.13 GPa, r_0 = 3.3 μm, s = 1 μm, ϕ_1 = 23.2 % and τ_a/τ_c = 1.195. ϕ_1 was calculated by Eq. (33) with $R_m = 2r_0$.

Recent results

We are presenting some recent results reached in the study of a BMI resin and unsized high-tenacity (TUS) and high-modulus (MUS) carbon fibers. The fibers have an imperfect graphitic structure with many oxygen groups and a low hydrogen content which enable the chemical bonding of modifiers on the CF surface. The selected modifiers can react with BMI resins. Based on the classical physico-chemical description of the carbon fiber surface, the reactions used in surface modification are Diels-Alder cycloaddition and nucleophilic substitution.

Diels-Alder reaction is typically carried out with conjugated dienes and olefins containing electron-withdrawing groups. The reaction does not occur if the diene structure is a part of a perfectly conjugated system such as benzene, but occurs if the diene is a part of polyconjugated ring systems [20]. In this case π electrons are more weakly bonded than in benzene, so that they can react with dienophiles such as tetracyanoethylene (TCNE), maleic anhydride (MA) and citraconic anhydride (CA).

The presence of hydroxy groups in the imperfect graphitic systems of carbon fibers can be used for their surface modification with bases such as ammonia and amines. The exchange reaction of OH group for amino group is not easy with phenol, but proceeds easily in polycyclic compounds containing more than four rings [21].

Both types of the reactions occur in the case of high-tenacity (TUS) and high-modulus (MUS) unsized fibers, but the different types and structures of MUS and TUS have a certain influence on the conversion. The MUS fiber reveals a higher order in microcrystalline graphitic structure than TUS due to the higher heat treatment temperature used in the fiber production starting from PAN precursors. We found by ESCA analysis that MUS fibers react with NH_3 reaching higher conversions of nucleophilic substitions than TUS under equal conditions.

Adhesion evaluation was done by the SFC test. Due to the low strain-at-break of BMI, all the adhesion tests were done using the coaxial SFC method. The fibers coated with BMI were embedded after a first cure in epoxy Ciba LY 556 + HT 972 and cured at 140 °C. Table 4 summarizes the results of l_c at room temperature in relation to the type of the treatment, both in the case of TUS and MUS fibers.

We can see that the shortest critical length l_c was obtained in the case of TUS treated with ammonia at room temperature (l_c = 0.39 mm). Quite similar results are obtained with MUS fibers (minimum l_c= 0.37 with NH_3), while the Diels-Alder treatment in this case gave higher l_c.

Table 4 Micromechanical results of T800 and M40 fibers/BMI/epoxy LY556-HT972 systems with different fiber treatments

Fiber	Sample	Reactant(s)[a]	Time/temperature (h/°C)	l_c (mm)
T800	101	ammonia	120/room	0.39
	66	ammonia	9/130	0.50
	112	ammonia	9/140	0.44
	3	40% MA+1% DCP	6/150	0.43
	30	50% MA+1% DCP	12/150	0.53
	62	50% MA	6/150	0.43
	31	CA + DCP	12/150	0.49
	0	as received (unsized or sized)	-	0.43
M40	88-108	ammonia solution	120/room	0.41
	11B-19B-	ammonia vapor	9/130	0.45-0.49
	74	ammonia vapor	15/130	0.37
	10C	TCNE	7/room	0.65
	5B	100% AA + BPO	10/80	0.55
	17B	EDTA + DCP	5/140	0.57
	13B	as received unsized	-	0.75
	1B	extracted	-	0.65
	75-2B-20B -	as received epoxy-sized	-	0.65

[a] MA maleic anhydride, DCP dicumyl peroxide, TCNE tetracyanoethylene, CA citraconic anhydride, AA acrylamide, BPO dibenzoyl peroxide

Figure 16 shows the dependence of l_C versus the nitrogen surface content of the treated TUS and MUS fibers. Nitrogen is not present in the MUS surface and therefore its presence is exclusively due to the bound ammonia molecules. It can be observed that in our cases, ammonia treatment is the most efficient way for increasing the adhesion between BMI and fiber. In fact, treatment with other nitrogen-containing compounds of higher MW have given higher l_C.

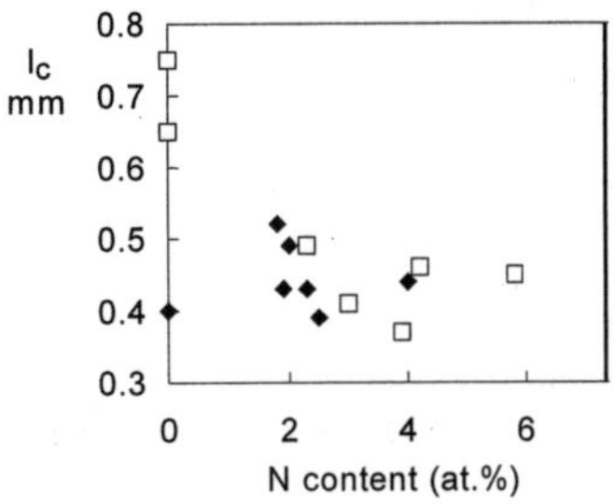

Fig. 16 Critical length measured on treated fibers with different surface nitrogen concentrations (□ M40, ◆ T800)

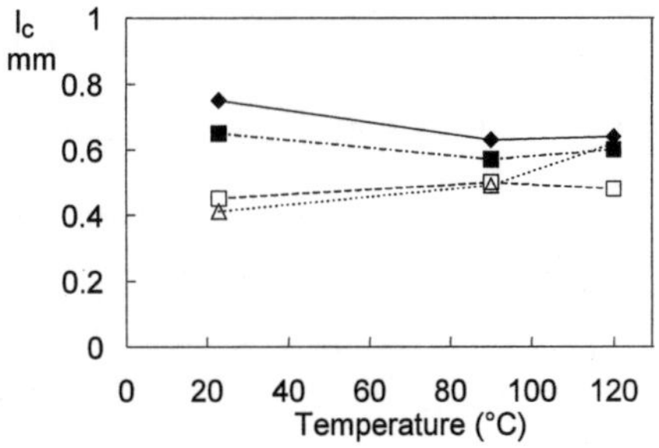

Fig. 17 Critical length of ammonia treated M40 fibers as function of temperature (◆ MUS, ■ MEBS, □ M40-11B, Δ M40-88/108)

Various tests were done to check the dependence of l_c versus T between room temperature and 120 °C. This upper limit is due to the low modulus of the embedding epoxy resin, which makes difficult to compare the results obtained at different temperatures. On increasing the test temperature, l_c increases and τ decreases.
Figure 17 shows the behaviour of the NH_3-treated MUS - BMI systems in dependence on temperature: it is interesting to observe that while no remarkable variation of adhesion is detected in TUS fibers after the treatment, MUS as such and even a commercial epoxy-sized M40 fiber (MEBS) appear to be characterized by a lower adhesion to BMI than ammonia-modified MUS.
Preparation of fiber tows impregnated with a BMI 5250-2 resin dissolved in acetone (50%) can be done by eliminating the solvent and curing the resin. The tows were submitted to tensile test at different temperatures. While the break load does not change on increasing temperature, due to the carbon fiber presence, the pull-out length distribution of the broken fibers can be studied on tows tested at different temperatures.
The length distribution at 20 and 250 °C is wide and the average length increases notably from 17.8 to 34 μm. At the moment, no reliable calculation can be done on these data to obtain the adhesion shear resistance.
At the highest temperature (250 °C), the radial stress given by Eq. 25 should be positive (tensile). Adhesion under this condition (250 °C) could be only due to chemical bonds existing at the interface or simply to the viscous drag motion between fiber and polymer network if the interface chemical bonds are at least partially effective.

REFERENCES

1. M.R. Piggot (Ed.), *Interfaces in Composites*, Elsevier 1991.
2. J.P. Favre, M.H. Auvray, P. Cheneau-Henry, C. Galiotis, C. Vlattas, A. Paipetis, M. Pegoraro, F. Severini, L. Di Landro, L. Yuan, *Polym. Compos.* 17, 936 (1996).
3. L. Di Landro, M. Pegoraro, *Composites* 27A, 847 (1996).
4. L. Di Landro, M. Pegoraro, *Chim. Ind*, 70(6), 86 (1988).
5. L. Di Landro, M. Pegoraro, *J. Mater. Sci.* 22, 1980 (1987).
6. M. Pegoraro, L. Di Landro, *Makromol. Chem., Macromol. Symp.* 70/71, 193 (1993).
7. A. Kelly, *Strong Solids*, Clarendon Press, Oxford 1973.
8. H.L. Cox, *Br. J. Appl. Phys.* 3, 72 (1952).
9. W. Weibull, *J. Appl. Mech. (ASME)* 18, 293 (1951).
10. A. Kelly, W.R. Tyson, *J. Mech. Phys. Solids* 13, 329 (1965).
11. C. Galiotis, in *Interfaces in Composites* (M.R. Piggot, Ed.), Elsevier 1991, p. 125.
12. A. Paipetis, C. Galiotis, *Composites* 27A, 755 (1996).
13. BREU CT 91-503, *Interface Contribution to the Temperature Dependent Properties of Carbon Fiber Reinforced BMI Composites*. Report 1993.
14. A.T. DiBenedetto, in *Interfaces in Composites* (M.R. Piggot, Ed.), Elsevier 1991, p. 103.
15. J.A. Nairn, *Polym. Compos.* 6 ,123 (1985).
16. L. Di Landro, A.T. DiBenedetto, J. Groeger, *Polym. Compos.* 9, 209 (1988).
17. F. Crescenti, C. Salvetti, M.S. Thesis, Politecnico di Milano, 1994.
18. S. Schutt, Ph.D. Thesis, Inst. für Kunststoffverarbeitung (IKV) and Politecnico di Milano, 1995.
19. J.P. Favre, D. Jacques, *J. Mater. Sci.* 25, 1373 (1990).
20. J. March, *Advanced Organic Chemistry*, 4th ed., J. Wiley 1992, p. 841.
21. L.F. Fieser, *Advanced Organic Chemistry*, Reinhold 1961, p. 889.

Collective Dynamics of Terminally Anchored Polymer Chains

Spiros H. Anastasiadis†*

Foundation for Research and Technology-Hellas, Institute of Electronic Structure and Laser, P.O. Box 1527, 71110 Heraklion, Crete, Greece

Abstract: Thermal fluctuations of the segment density profile of a polymer brush were probed by dynamic light scattering in the evanescent wave configuration; the time correlation functions of concentration fluctuations with wavevector q were measured. It is found that there is a preferred wavelength of the order of the brush thickness, L_0, of these fluctuations with a concurrent slowing down of their thermal decay rate. A theory is presented for the small-amplitude deformation of the free surface of a parabolic brush in solution; a maximum in the respective structure factor of the concentration fluctuations is predicted at $q^* L_0 \sim O(1)$, in agreement with the experiment.

INTRODUCTION

Polymer adsorption has numerous consequences in disciplines from biology to tribology and technological applications like surface/interface modification, adhesion and wetting, and stabilization of colloids. Selective adsorption from solution via attaching the ends of polymer chains onto a solid surface can lead to a grafted layer due to chain extension away from the surface, a configuration termed a polymer brush (Ref. 1); this configuration of stretched tethered chains is a model structure for a large variety of confined polymer systems (Ref. 2). Most of the theoretical and experimental work, so far, have addressed the static structure of tethered chains (Refs. 2-7), whereas the absence of direct experimental observation of their dynamic structure together with the strong relevant theoretical predictions available (Refs. 3,8) invite new experimental efforts in this active field (Ref. 5).

In this work, the time correlation functions of composition fluctuations at thermal equilibrium are measured in poly(ethylene oxide)-*block*-polystyrene block copolymers, PEO-PS, adsorbed from toluene solution onto a glass surface with their short PEO blocks (Ref. 9). The evanescent electromagnetic wave, produced under total reflection of a laser beam, is utilized as the incident beam in the scattering experiment; the time correlation functions $C(q,t)$ of concentration fluctuations with wavevector q are obtained taking advantage of the broad time range (10^{-7}-10^{3} s) of photon correlation spectroscopy (PCS) (Ref. 10). Besides the applicability of the evanescent wave dynamic light scattering (EWDLS) to dynamics study of tethered chains (Ref. 9), the experiments demonstrate that long-lived fluctuations possess a finite wavelength of $O(L_0)$ as indicated by the apparent maximum in the scattering intensity at a finite wavevector, q^*. These observations indicate that the fluctuations observed may be associated with a small-amplitude deformation of the free surface of the brush (Ref. 11). We

† In collaboration with G. Fytas, D. Vlassopoulos, A. Likhtman, A. N. Semenov, C. Toprakcioglu, J. Li, B. Factor, R. Seghrouchni, W. Theobald.

* Also at University of Crete, Physics Department, 711 03 Heraklion Crete, Greece.

 CCC 1022-1360/99/$ 17.50+.50/0

present a theory (Ref. 12) for the calculation of the excess Helmholtz energy of a small sinusoidal deformation of the free surface of a parabolic brush in solution and show that there is a maximum in the respective structure factor of the concentration fluctuations at $q^* L_0 \sim O(1)$; this is due to a balance between elastic and concentration gradient contributions to the Helmholtz energy functional.

EXPERIMENTAL

The detailed material and surface characterization is essential for the interpretation of the results obtained from this new experimental technique. The PEO-PS model system in toluene was selected due to the extensive information on the adsorbed amount, kinetics of adsorption, and segment density profiles obtained from ellipsometry, neutron reflectivity, and surface balance experiments (Ref. 7). The characteristics of the PEO-PS diblocks and the brush-like grafted layers are given in Table 1. The brush is formed by the stretched PS block dangling in the solution; the anchoring density is $\sigma = s^{-2}$ (chains nm^{-2}), with s denoting the interanchor spacing. The equilibrium thickness L_0 of the grafted PS monolayer corresponds to a parabolic density profile normal to the surface (Refs. 1a, 7c). The adsorption conditions of Ref. 7 were reproduced, and the copolymer concentration for maximum adsorption A^* was used; equilibrium was confirmed by the stability of the scattered intensity. The temperature throughout the experiments was 23 °C.

Table 1. Molecular and brush characteristics of the PEO-PS in toluene (Ref. 7)

Code	M_W	N_{PS}	N_{PEO}	A^* mg m^{-2}	L_0 nm	s nm
PEO-PS/80	80 000	730	90	2.9	45±3	7±0.5
PEO-PS/150	150 000	1420	51	1.6	70±5	2±0.5
PEO-PS/184	184 000	1700	167	2.6	80±5	11±0.5
PEO-PS/500	502 000	4790	91	1.7	130±10	23±1

The experimental set-up is schematically presented in Fig. 1. Note that few attempts to measure dynamic properties of strong scatterers (e.g., colloidal particles) close to a wall by EWDLS have been reported (Ref. 13). For weak scattering, however, careful optical alignment to minimize parasitic scattering and specific control experiments to ensure homodyne detection for both static and dynamic measurements are crucial for the quality of the experimental functions (Ref. 10) and in order to verify that the standard assertions used in bulk solution measurements by PCS are accurate for the present EWDLS.

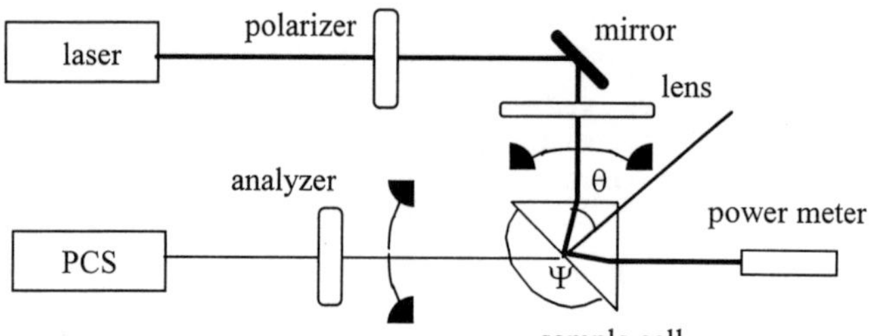

Fig. 1. Schematic of the evanescent wavedynamic light scattering instrumentation (Ref. 9). The light scattering cell is a right-angled high-refractive-index prism to which a semicylindrical cell is epoxy-glued.

The evanescent wave geometry is shown in Fig. 2. For total reflection [$\theta > \theta_c = \arcsin(n_2 / n_1)$], the evanescent wave propagating parallel to the surface, with an electric field exponentially decaying away from the surface and with penetration depth ζ, is scattered by the concentration fluctuations of wavevector $\mathbf{q}$ ($|\mathbf{q}| = 2k_i \sin(\Psi / 2)$) in the adsorbed polymer layer.

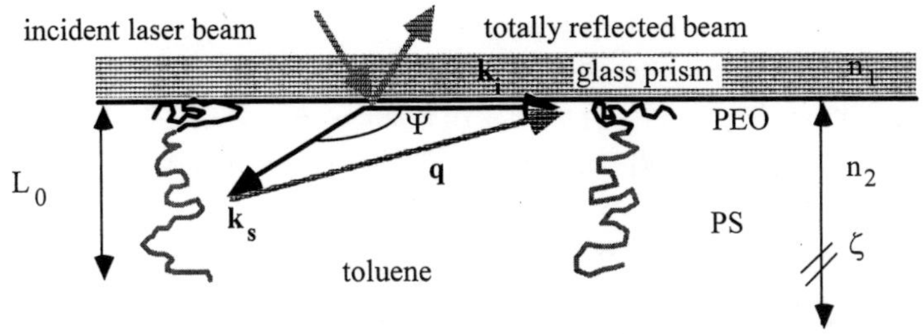

Fig. 2. The evanescent wave geometry versus the polymer brush (Ref. 9).

The measured intensity autocorrelation function $G(q,t)$ leads to the computation of the desired $|C(q,t)|^2 = \left(G(q,t) - A\right) / f$, where A is the baseline at sufficiently long delay times and f is a coherence (instrumental) factor.

RESULTS AND DISCUSSION

Figure 3 shows $C(q,t)$ for the tethered PS layers. The featureless $C(q,t)$ from pure toluene, as expected, becomes a well defined decay function after a 7-h counting time for PEO-PS/500 at $q = 0.034$ nm^{-1}. This surface relaxation function clearly reflects the dynamics of the grafted layer and not fluctuations due to the translational motion of free polymer chains (solid line) measured in the "bulk"geometry obtained in situ for $\theta = \theta_C + 5°$, where the incident light penetrates into the solution as a refracted beam. $C(q,t)$ is well represented by a single exponential function

$$C(q,t) = a(q)\exp[-\Gamma(q)t] \tag{1}$$

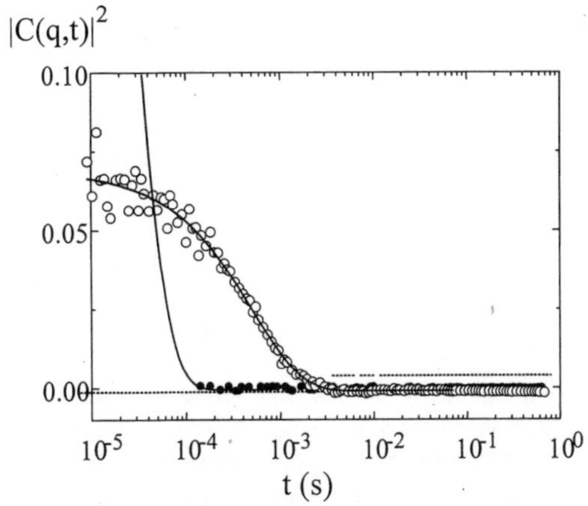

Fig. 3. $C(q,t)$ for toluene (●) and for the PEO-PS/500 brush in toluene (0.4 mg/ml and 23 °C) at q=0.034 nm^{-1} (O), and the distinctly different correlation function from the "bulk" PEO-PS/500 solution (—).

where $\Gamma(q)$ is the thermal decay rate of the composition fluctuations in the layer. The value of the relaxation amplitude $a(q)$ shows that about 25 % of the total scattering intensity, I_{tot}, is due to the probed polymer layer mode. Both $C(q,t)$ and the process intensity $I(q) = a(q)I_{tot}$ depend on the scattering wavevector q (Ref. 9), where $I(q)$ increases with q (Fig. 4), implying strong suppression of the low wavevector (long wavelength) fluctuations, whereas $\Gamma(q)$ for all but the longest PEO-PS chain is virtually insensitive to q; for PEO-PS/500, a slowing down of the rate is observed at high wavevectors. The intensity data for the four different layers superimpose well when $I(q)$ is divided by A^* and q is normalized to L_0 (Fig. 4). Actually, the reduced intensities $I(q)/A^*$ for all layers are found to exhibit a $(qL_0)^2$ dependence in the lower experimental qL_0 range (~1-4; Fig. 4); for the longest PEO-PS/500 layer and at higher qL_0 values, $I(q)$ exhibits a higher rate of increase with q with indications of a broad maximum at the edge of the highest accessible q range.

Theory predicts the end-to-end chain relaxation for a grafted brush to depend on N and on the anchoring density σ (Ref. 8). Considering mean-squared fluctuations of the chain free end in the perpendicular direction proportional to L_0^2, a Rouse-like diffusion led to the characteristic q-independent relaxation rate

$$\Gamma \propto w(\sigma)\sigma^{-\beta}N^{-3} \tag{2}$$

with $w(\sigma)$ denoting the microscopic mobility (can depend on σ), and β = 2/3 for free-draining chains or β = 1 when hydrodynamic interactions are taken into account (Ref. 8b). Molecular dynamics and Monte-Carlo simulations (Ref. 14) were consistent with the scaling prediction of Eq. 2.

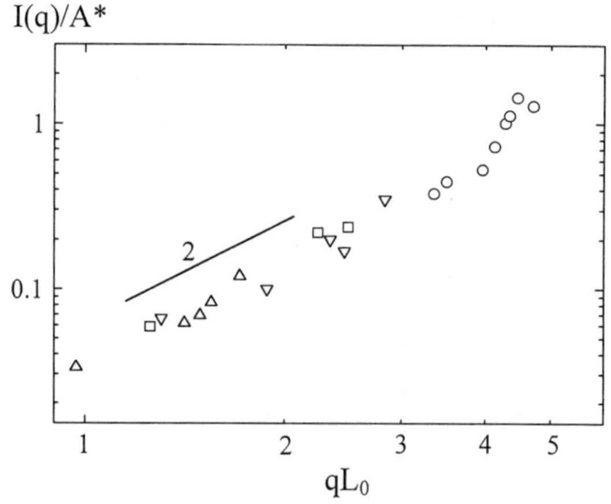

Fig. 4. Variation of the intensity normalized to the adsorbed mass per area, A^*, for all four brushes with the product of the wavevector times the layer thickness: Δ PEO-PS/80, □ PEO-PS/150, ∇ PEO-PS/184, O PEO-PS/500. (The line has slope 2.)

Dynamic light scattering probes collective dynamics. Thus, the experimental $\Gamma(q)$ shows the effect of thermodynamics, i.e., of $I(q)$. Therefore, $\Gamma(q)$ is q-independent for the three shorter brushes and shows a slowing down for PEO-PS/500 at high wavevectors. The N-

dependence of $\Gamma(q)$ at low qL_0 (Eq. 2) is revealed when the dependence of the microscopic mobility on concentration (and thus on σ) is taken into account.

Recently, small-amplitude deformation of the free surface of an incompressible, molten „step function“ polymer brush was analyzed and the Helmholtz energy, F, relative to that of the undeformed flat layer was computed (Ref. 11). Both lateral and perpendicular chain displacements were considered and the correlation function $P(q)$ of height fluctuations at the surface exhibited a maximum at a certain value of $q^* L_0$ since both long- and short-wavelength fluctuations are suppressed at the entropic and interfacial tension (γ) expense, respectively. This led to a preferred wavelength (parallel to the polymer/air interface) $2\pi / q^*$ of the most probable thermally-excited surface modes. The value of $q^* L_0$ depends on the relative contributions to F of the γ-term at high q versus that due to chain stretching. Relaxing the assumption that all free chain ends reside at the top layer surface (Ref. 11b) leads to only a slight modification.

We have extended (Ref. 12) the earlier work for the case of small-amplitude deformation of the free surface of a parabolic brush in solution (Fig. 5) by calculating the contributions to the Helmholtz energy of the elastic term due to chain stretching and of a gradient contribution to the non-constant concentration in both x and z directions, where straight chain trajectories were assumed.

In each wedge (Fig. 5), the molecular field is a parabolic function of the distance from the junction point along the chain, $h(x)$, i.e., $U = A - B[l(x)]^2$, where $l^2(x) = z^2(1+f^2(x))$, $f(x)$ is the tangent of the angle between the chain trajectory and the perpendicular to the plane z = 0 (z is the axis perpendicular to the grafting plane), $h(x)$ is the height of the brush, $\{x + \mathrm{f}(x)h(x), h(x)\}$ are the coordinates of the free chain end, and $\phi(x,z) = -U/\omega$ is the density of polymer chains, ω being the excluded volume parameter. The polymer density should be zero at the brush edge, $z = h(x)$, i.e., $\phi(h(x)) = A(x) - Bh^2(x)(1 + f^2(x)) = 0$, whereas the total amount of polymer inside one wedge must be equal to the number of grafted monomer units, $N\sigma$: $\int_{\text{wedge}} \phi(x,z)\mathrm{d}z = N\sigma$.

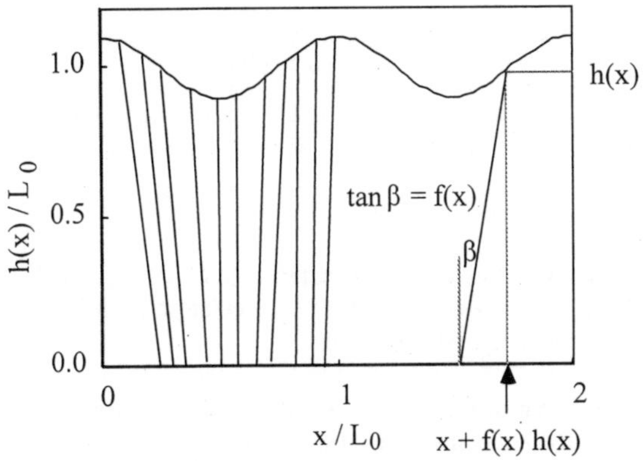

Fig. 5. Schematic diagram of the geometry of a deformed polymer brush in solution.

Since there are three unknown functions and only two equations, one needs to predefine one of the unknowns; it is assumed that the function $f(x) = -(8\varepsilon / (qL_o))\sin(qx)$. Then, one gets $h(x) = L_o\left(1 + \varepsilon\cos(qx) + O(\varepsilon^2)\right)$ with $L_o = (24/\pi^2)^{1/3} N(\sigma\omega\alpha^2)^{1/3}$, and $\alpha = b/\sqrt{6}$ where b is the statistical segment length. The total elastic Helmholtz energy per unit area is then

$$\frac{F_{\text{elastic}}}{F_0} = 1 + \varepsilon^2\left(\frac{7}{2} + \frac{160}{9q^2h_0^2}\right) \tag{3}$$

with $F_o = (9/10)\left(\pi^2/24\right)^{1/3} N\sigma(\sigma\omega/\alpha)^{2/3}$ is the Helmholtz energy of the undeformed brush. The gradient term contribution to the Helmholtz energy per surface area is calculated as

$$F_{\text{gradient}} = \frac{\alpha^2}{4}\frac{q}{2\pi}\int_0^h \mathrm{d}z \int_0^{2\pi/q} \mathrm{d}x \frac{(\nabla\phi)^2}{\phi} =$$

$$= \frac{\pi^2 L_0}{16N^2\omega}\left\{\left(1 + \frac{1}{2}\ln\frac{2}{\delta}\right)\left[1 + \frac{64\varepsilon^2}{3(qL_0)^2} + \frac{3\varepsilon^2}{2}\right] + \varepsilon^2\frac{(qL_0)^2}{4}\ln\frac{2}{\delta} - 2\varepsilon^2[-\ln(2\delta) - 1]\right\} \tag{4}$$

where $\delta = \xi / h_0 = N^{2/3}(a/h_0)^{4/3} = N^{-2/3}[a/(\sigma\omega)]^{4/9} << 1$ with $\xi \cong \alpha^{4/3}N^{2/3}/L_o^{-1/3}$ being the terminal correlation length (Ref. 16).

The final result can be presented in the form $F = F_0 + \frac{1}{2}(\Delta\Phi)^2[S(q)]^{-1}$ with $\Phi = \int_0^h \phi(x,z)\mathrm{d}z$ and $\Delta\Phi \equiv \Phi - \Phi_0 = 3N\sigma\varepsilon$, where the structure factor of the fluctuations of total concentration in the brush is

$$[S(q)]^{-1} = \frac{1}{5}\left(\frac{\pi^2}{24}\right)^{1/3}\left(\frac{\omega\sigma}{\alpha}\right)^{2/3}\frac{1}{N\sigma}\left\{\frac{7}{2} + \frac{160}{9(qL_o)^2} + \frac{5(3\pi)^{2/3}\ln(2/\delta)}{72}\left(\frac{\alpha}{\omega\sigma}\right)^{4/3}\frac{(qh_0)^2}{N^2}\right\} \tag{6}$$

Thus, theory predicts a preferable q^* for maximum fluctuations at

$$q^* L_0 = \left[\frac{256}{(3\pi)^{2/3}\ln(2/\delta)}\right]^{1/4} N^{1/2}\left(\frac{\sigma\omega}{a}\right)^{1/3} \tag{7a}$$

i.e., at

$$q^* = \frac{2}{3^{1/2}[\ln(2/\delta)]^{1/4}}\frac{1}{\alpha N^{1/2}} \tag{7b}$$

Actually, the predicted value for q^* for the PEO-PS/500 copolymer is $q^* \cong 0.039\ \text{nm}^{-1}$, right near the edge of our experimental q range. At low wavevectors,

$$S(qL_o \to 0) = 0.38\left(\frac{\omega\sigma}{\alpha}\right)^{-2/3} N\sigma(qL_o)^2 \tag{8}$$

in agreement with the behavior of the experimental data (Fig. 4). Note that $A^* = N\sigma$. Currently, the theory is being extended for the calculation of the full dynamic structure factor $S(q,t)$, i.e., including its temporal dependence.

Summarizing, the evanescent wave dynamic light scattering technique allowed the investigation of the dynamic structure of terminally anchored PS layers, which revealed the presence of long-lived thermally induced layer fluctuations with wavelength $2\pi/q^* \cong 1.3L_0$ indicating strong surface coverage effects on their thermal decay rate. The identification of the precise mechanism of the layer fluctuations would highly benefit from a theoretical calculation of the full structure factor $S(q,t)$. This will promote fundamental studies of polymers near surfaces and provide information on chain relaxation and brush friction, relevant to technological applications.

ACKNOWLEDGEMENTS

Part of this research was sponsored by the European Union (International Scientific Cooperation, Contract ISC*CT93-0951) and by NATO's Scientific Affairs Division (Science for Stability Programme).

REFERENCES

1. (a) S. T. Milner, *Science* **251**, 905 (1991); (b) R. Yerushalmi-Rozen and J. Klein, *Phys. World* 1995 (August), 30.
2. A. Halperin, M. Tirrell, T.P. Lodge, *Adv. Polym. Sci.* **100**, 31, (1992).
3. (a) P. G. de Gennes, *Adv. Colloid Interface Sci.* **27**, 189 (1987); (b) P. G. de Gennes, *C.R. Acad. Sci., Ser. II* **302**, 765 (1986).
4. H. E. Johnson and S. Granick, *Science* **255**, 966 (1992).
5. J. T. Koberstein (Ed.), *Polymer Surfaces and Interfaces*, *MRS Bull.* **21**, 16 (1996).
6. B. Farago, M. Monkenbusch, D. Richter, J. S. Huang, L. J. Fetters, A. P. Gast, *Phys. Rev. Lett.* **71**, 1015 (1993).
7. (a) H. J. Taunton, C. Toprakcioglu, L. J. Fetters, J. Klein, *Macromolecules* **23**, 571, (1990); (b) H. Motschmann, M. Stamm, C. Toprakcioglu, *Macromolecules* **24**, 3691 (1991); (c) J. B. Field, C. Toprakcioglu, R. C. Ball, H. B. Stanley, L. Dai, W. Barford, *Macromolecules* **25**, 434, (1992).
8. (a) A. Halperin and S. Alexander, *Europhys. Lett.* **6**, 329 (1988); (b) L. I. Klushin and A. M. Skvortsov, *Macromolecules* **24**, 1549 (1991).
9. G. Fytas, S. H. Anastasiadis, R. Seghrouchni, D. Vlassopoulos, J. Li, B. J. Factor, W. Theobald, and C. Toprakcioglu, *Science* **274**, 2041 (1996)
10. G. Fytas and G. Meier in *Dynamic Light Scattering. The Method and Some Applications* (W. Brown, Ed.), Oxford Clarendon Press, London, 1993.
11. (a) G. H. Fredrickson, A. Ajdari, L. Leibler, J. P. Carton, *Macromolecules* **25**, 2882 (1992); (b) H. W. Xi and S. T. Milner, *Macromolecules* **29**, 4772 (1996); (c) F. J. Solis and G. T. Pickett, *Macromolecules* **28**, 4307 (1995); (d) D. Long, A. Ajdari, L. Leibler, *Langmuir* **12**, 1675 (1996).
12. A. E. Likhtman, S. H. Anastasiadis, and A. N. Semenov, *Macromol. Theory Simul.*, submitted .
13. K. H. Lan, N. Ostrowsky, D. Sornette, *Phys. Rev. Lett.* **57**, 17 (1986).
14. (a) M. Murat and G. S. Grest, *Macromolecules* **22**, 4054 (1989); (b) P. Y. Lai and K. Binder, *J. Chem. Phys.* **95**, 9288 (1991).
15. S. Milner, *Macromolecules* **21**, 2610 (1988).
16. A. N. Semenov, *Macromolecules* **25**, 2273 (1992).

Modification of Surfaces by Covalent Attachment of Polymer Micelles

Jungseok Hahn and Stephen E. Webber*

Department of Chemistry and Biochemistry and Center for Polymer Research
The University of Texas at Austin, Austin, Texas 78712, USA

Abstract: Polymer micelles composed of polystyrene-*block*-poly(methacrylic acid) have been chemically attached to amino-modified SiO_2 surfaces by carbodiimide coupling. The density of coverage is very high (ca. 3 x 10^{10} micelles/cm^2) and fairly uniform. The diameter of the micelles is ca. 50 nm and the distribution of micelle sizes is essentially undisturbed by this coupling reaction. These modified surfaces might be of interest in surface modification because of the very high density of reactive groups that are associated with each micelle.

BACKGROUND

Diblock copolymers can exhibit self-assembly capabilities due to different solvent selectivity of each block (Ref. 1). As a result of this, spherical micelle structures can be assembled with the micelle core formed by the poorly solvated block and the micelle corona from the well-solvated block. Amphiphilic block copolymers in water have received attention lately because of their scientific interest and potentially useful industrial applications (Ref. 2). For example, the release kinetics of hydrophobic chemicals from water-soluble micelles has been studied as a model of hydrophobic drug delivery in medicine (Ref. 3) and the detergent-like behavior of these kinds of micelles has been used for removal of specific organic pollutants (Ref. 4).

If water or aqueous buffers dissolve the polyelectrolyte block and strongly precipitate the remaining nonpolar blocks, the micelles exhibit strong segregation of the hydrophobic core and polyelectrolyte corona (Ref. 5). In the case of polystyrene, the high glass transition temperature assures that the core will be rigid and dense (Ref. 5b). Most micellization processes in aqueous solvents are not carried out under equilibrium conditions and each step of micelle preparation is significant for the properties of resulting micelles once mass exchange between unimers and micelles ceases (Ref 5d) or if none of polymers exist as unimers. In some cases amphiphilic block copolymers can spontaneously form micelles (Ref. 5a). Stepwise dialysis from organic solvent-water mixtures to pure water or buffer is often employed to prepare these micelles. After a certain composition of solvent in the dialysis is achieved and there is no further change in the association number but the hydrodynamic

 CCC 1022-1360/99/$ 17.50+.50/0

diameter of a micelle, which is primarily determined by the stretching of the corona, is highly sensitive to solvent composition (Ref. 5b,f,h).

The objective of these studies is the chemical attachment of polymeric micelles onto solid surfaces. Multiblock copolymers on surfaces and interfaces have been investigated by theoretical calculations and experimental model systems and industrial applications have been explored in fields as diverse as aerospace and electronics (Ref. 6). The efforts in this field have contributed to technological developments in the stabilization of dispersed particles, characterization methods, and surface modification. An important aspect of understanding the behavior of polymeric micelles on surfaces lies in characterization of their fixation process onto surfaces. The research described in this report provides techniques and chemistry associated with covalent attachment of water-soluble polymeric micelles assembled from polystyrene-*block*-poly(methacrylic acid) (PS-PMA) on quartz surfaces. Similar studies have been carried out previously on absorption of micelles on polystyrene films (Ref. 7), surface micellization of block copolymers on quartz or liquid surfaces (Ref. 8) and micelle attachment on silicon nitride surfaces (Ref. 9).

EXPERIMENTAL

Copolymers

The detailed synthesis and characterization of the polystyrene-*block*-poly(methacrylic acid) copolymer has been reported (Ref. 10). The PS-PMA copolymer was prepared by a living anionic polymerization in tetrahydrofuran at -78 °C. The weight fraction of PS block (W_{PS}) is 0.64 from NMR analysis and the weight average molecular weights are 2.83×10^4 g/mol for PS block and 1.54×10^4 g/mol for PMA block, as determined by GPC and NMR analyses. The molecular weight of the PS-PMA micelle measured by static light scattering is 1.25×10^7 g/mol which corresponds to an aggregation number of 286 (Ref. 11).

Preparation of Micelles

1,4-Dioxane (99.0 %) was used as received from EM science. The PS-PMA copolymer was dissolved in a dioxane/deionized water mixture (80 volume % dioxane) to produce micelles directly. The solution underwent stepwise dialysis against a series of dioxane/water mixtures with 60, 40, and 20 volume % dioxane. A fifty-fold volume excess solvent mixture was used at each step for 4 h and typically a 20 % dioxane step was performed twice. PVDF filters (0.45 mm pore size, Gelman) were used to filter the micellar solution although there is no visible aggregation.

Preparation of Reactive Surface

Fused quartz discs (1" x 1/16") were obtained from Quartz Plus, Inc. The quartz surfaces were cleaned by dipping the discs in a series of acid solutions: 10% HNO_3 for 1-2 min, 50/50 concentrated HCl/methanol for 1-2 min, and concentrated H_2SO_4 for 30-60 s. The surfaces were rinsed thoroughly with deionized water after each dipping and air-dried.

A 5 vol.-% (3-aminopropyl)trimethoxysilane (APTMS, b.p. 91-92 °C/15 mm, Aldrich) solution was prepared by mixing APTMS with toluene (Fisher Scientific) which was passed through an alumina column to remove water. The vapor phase silanization reaction was performed by placing the acid-cleaned quartz discs (DS1, DS2) in a stationary open holder located above refluxing 5 vol.-% APTMS in toluene solution at 97 °C for 6 h. The amino-activated quartz surfaces were washed with toluene, air-dried, and stored in a vacuum desiccator. A xylene solution of {3-[(2-aminoethyl)amino]propyl}trimethoxysilane (b.p. = 140 °C/15 mm) which has two active amino groups was also tried, but it failed since there was not enough vapor pressure of silane at the boiling point of xylene. In order to improve the uniformity of amino groups on surfaces, a device for rotating the quartz discs (DR1, DR2) during the vapor-phase reaction replaced the stationary holder.

The reaction mechanism for this process is well known: hydrolysis of alkoxy groups of APTMS occurs either by absorbed H_2O on substrate surfaces or with OH groups of the surfaces. In the former case, the resulting silanols can form covalent linkages with the substrate by losing water. However, these silanols may condense with themselves to form oligomers on the surface. Therefore, the dryness of the equipment and the surfaces is a critical factor of the reaction. The surfaces and the apparatus including the rotary disc holder were dried in an oven at 120 °C for 30 min before the surface activating reaction.

Micelle Attachment

The PS-PMA micelles were attached to the aminated surface by the reaction of the carboxyl groups on the micelle coronas using a carbodiimide reagent. A dilute 1-(3-dimethylaminopropyl)-3-ethylcarbodiimide methiodide (Aldrich) solution in a 80-vol.-% dioxane/water mixture was mixed with a micelle solution and the activated quartz disc was then immersed into the carbodiimide and micelle mixture (12.0 mL) for 6 h at ambient temperature. The samples denoted 1065-10, 1065-15, 1065-20, and 1065-25 were prepared from 1.0, 1.5, 2.0, and 2.5 mmol of the carbodiimide in the reaction mixtures, respectively. The concentration of micelles was 1.065 mg/mL, which is equivalent to 4.45 mmol of carboxylic acid groups. The surface micelle samples were rinsed with 80 % dioxane/water mixture and air-dried completely before measurements.

Characterization of Surface

Water Contact Angle Measurement

Water contact angles were measured with NRL C.A. Goniometer (Ramé-hart, Inc.). Deionized water droplets (maximum of 100 μL) were used in all measurements at room temperature. The final readings were taken approximately 1 min but no more than 3 min after setting the water droplet on the level surface. The surface was divided into several sectors and then at least 20 water contact angles were measured for a quartz surface.

X-Ray Photoelectron Spectroscopy

To study the availability of the amino groups on surfaces with XPS, the activated surfaces were submerged into a stirred 3.5 vol.-% trifluoroacetic anhydride (TFAA, 99+ %, Aldrich) solution in dichloromethane (EM Science) at room temperature for 6 h. TFAA can react readily with the amino groups and has a distinct fluorine peak in XPS, which allows an estimate of the fraction of amino groups that are reactive towards carboxylic acids.

The nitrogen and fluorine composition of the amino surface activated with TFAA was measured by XPS with Physical Electronics PHI 5700 ESCA System Model 10-610. Since quartz is an insulator, the sample became charged during the experiment. The survey scan over 0 - 1400 eV was performed at a pass energy of 93.900 eV and an energy step of 0.25 eV. The monochromatic Al Kα X-rays were used at a power of 300 W. The base pressure was 3.5×10^{-10} Torr when the data were acquired. Several points on a surface were measured to obtain statistical results.

Scanning Electron Microscopy

The SEM samples were prepared by taping the micelle-attached surfaces on the pin mount for SEM (1 cm in diam., Ted Pella, Inc.). A touch of colloidal silver liquid (Ted Pella, Inc.) was applied on each surface in order to assure electrical contact. The surfaces were then coated with Au in a Ladd bench-top sputter at 2.5 kV and 20 mA for 60 s, which produced ca. 15 nm thickness of Au on the surfaces. Electron micrographs were acquired on a Philips Model SEM-515 scanning electron microscope with a Polaroid Camera at 25.3 kV with a beam size of 10 nm.

Atomic Force Microscopy

The micelles on the surface were also studied by AFM with Auto Probe cp AFM (Park Scientific). Ultralevers from Park Scientific were used as the cantilever and tip. The sample was taped on a stainless steel sample support (2 cm in diam.) which was placed on a piezoelectric scanner with a maximum scan range of 100 mm x 100 mm. The non-contact image was acquired while the displacement of the probe was used as the feedback in the constant-force mode.

RESULTS

SiO_2 Surface After Silane Reaction

Contact angle measurements and XPS are well-known analytical tools to monitor the uniformity and completeness of surface modification reactions. The surface was divided into several sectors and then at least 20 water contact angles were measured for a quartz surface. The nitrogen of APTMS on the surface can be easily detected by XPS. The survey spectrum of the aminated surface shows the very distinct characteristic O1s, C1s, Si1s, and Si2p peaks, but the N1s peak is small due to its low concentration. In our chemical treatment, carbon impurities were unavoidable, so the carbon composition is not a reliable measure of the surface composition. The Si and O peaks are mainly from the abundant Si and O in quartz, so the ratio of silicon to nitrogen or oxygen to nitrogen cannot be used to find the absolute density of amino groups on the surface. The ratio of the distinct N1s and F1s peaks was used to find the availability of amino groups after reacting TFAA with surface amino groups.

The results for the rotated (DR1, DR2) and unrotated (DS1, DS2) samples are summarized in Table 1.

Table 1. Water contact angles and XPS of quartz-supported PS-PMA micelles

	H_2O contact angle (°)		XPS
Sample[a]	acid-cleaned	+ aminosilane	F/N
DS1	23.1 ± 0.7	68.8 ± 3.6	0.60 ± 0.09
DS2	24.5 ± 0.2	64.9 ± 6.0	1.07 ± 0.18
DR1	24.5 ± 0.2[b]	71.3 ± 2.7[c]	1.16 ± 0.12
DR2	---	69.2 ± 3.0[c]	0.84 ± 0.10

[a] DS - a stationary disk during silanization, DR - a rotated disk; [b] average over both disks; [c] after TFAA treatment the contact angle was $(65.7 \pm 4.6)^o$

Rotation reduced the deviations in contact angle readings and composition ratio for a given sample and also the values for the two disks were in better agreement. We conclude that the rotary holder did improve the uniformity of the amino-surface. Although an excess of TFAA was reacted for quite some time, the composition ratios were still far short of the theoretical value. Therefore we conclude that only ca. one-third of amino groups on the surface may participate in the amide bond formation reaction when the micelles are attached onto the surface. The contact angle of the TFAA-treated surface is slightly smaller than the initial silane surface (Table 1). This is certainly not what you would expect if the surfaces were uniformly covered by CF_3COO groups.

Characterization of micelle-attached surface

After TFAA treatment, the PS-PMA micelles were chemically attached onto the amino-activated quartz surface as described earlier. Four different carbodiimide concentrations in a range suggested in previous study (Ref. 9) were tried to characterize the effect of carbodiimide concentration on the size and polydispersity of the micelles on surface.

A typical SEM picture at normal incidence for the sample surfaces and the corresponding size distribution is given in Figure 1. The attachment process used in this study is highly efficient. The surfaces were densely covered by the micelles and in general the coverages were uniform. However, for 1065-25, which has the highest carbodiimide concentration, the variation was larger (see Table 2). The statistics of particle size for each sample showed an almost perfect normal distribution (see Figure 2 for a typical example).

Figure 1. SEM image of quartz-supported PS-PMA micelles.

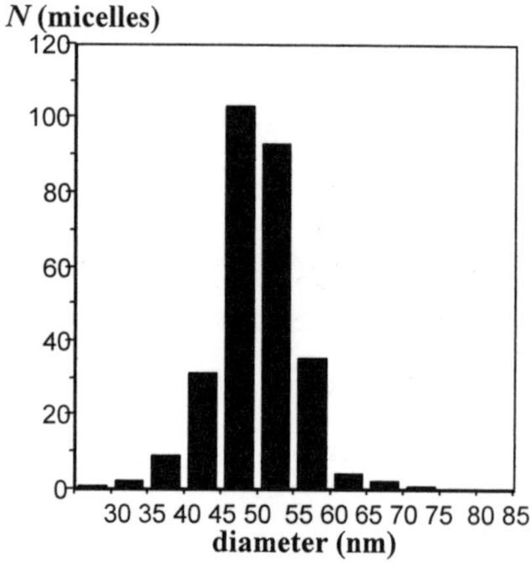

Figure 2. Size distribution of quartz-supported PS-PMA micelles. (Average = 50.1 ± 5.9 nm.)

It can be concluded from the variance that the polydispersity of the surface micelles is fairly small. The distribution analysis of SEM pictures is presented in Table 2. There is an increase in the surface density of micelles and the number of aggregates with carbodiimide concentration although the average size of the micelles was not dependent on the carbodiimide concentration. The aggregates are probably formed in the solution phase before the actual attachment reaction.

Table 2. Distributions in 1 μm^2 of SEM images of quartz-supported PS-PMA micelles

Sample	Average size nm	Number of micelles	Number of aggregates
1065-10	50.1 ± 5.9	281	4
1065-15	45.6 ± 5.4	313	2
1065-20	45.9 ± 5.9	346	11
1065-25A	51.7 ± 7.9	242	12
1065-25B	46.3 ± 6.2	378	16
1065-25C	45.1 ± 7.1	479	66
-25 series	46.7 ± 7.0	366	31

AFM was not successful in obtaining the height distribution of the micelles because the surface was too densely covered by PS-PMA micelles and a relatively large size AFM tip was used. Because the micelles on surface were presumed to be soft compared to other samples typically examined by AFM, a non-contact mode was used. A rough estimate of the height of the surface micelles was in the range of 10 - 20 nm.

The change in carbodiimide concentration did not influence the contact angle measurements of the micelle surface samples. The average contact angles for all samples was $(91.2 \pm 1.1)^{\circ}$ degrees, which was much higher than expected for a hydrophilic micelle corona (Ref. 7). We presume this is because of the excess carbodiimide on the micelle corona which reacts with the carboxylic acid to form a hydrophobic urea-like adduct. This adduct will cover that part of the micelle corona which was not used in the attachment reaction.

A rough calculation of the contact angle of the surface micelle may be carried out. Adamson gives the following equation for a composite surface (Ref. 12):

$$\cos \theta_{exp} = r \cos \theta_{true} = f_1 \cos \theta_1 + f_2 \cos \theta_2 \qquad (1)$$

where f_i is the fraction of area of type i, θ_i is the corresponding contact angle for i, and r is the roughness factor. Without consideration of r, Eq. 1 can be written

$$\cos \theta_{exp} = (1 - f_{mic}) \cos \theta_{amine} + f_{mic} \cos \theta_{mic} \quad (2)$$

Taking the average density of micelles on the surface as N and $\pi(d_{mic}/2)^2$ as the geometric factor

$$f_{mic} = N \pi (d_{mic}/2)^2 \quad (3)$$

The diameter of the micelles from QELS was ca. 54 nm and from the analysis of the SEM images a range from 45 to 52 nm is estimated. Combined with the value of N from Table 2, f_{mic} is calculated to be in the range 0.55 - 0.62. Since θ_{exp} and θ_{amine} are known (ca. 91.2° and 70.2°, respectively) and we have estimated f_{mic}, we solve Eq. 2. for $\cos\theta_{mic}$ and obtain θ_{mic} in the range 104 - 108°, which is very hydrophobic.

CONCLUSION

Chemical attachment of polystyrene-*block*-poly(methacrylic acid) micelles onto chemically modified quartz surface was accomplished by an amide-forming reaction with water-soluble carbodiimide. Water contact angles and XPS were measured to verify the siloxane modification reaction and to check the uniformity of resulting surface. The uniformity of amino-activated surface, which is essential to obtain homogeneous coverage of micelles, was improved by using a device for rotating the quartz discs during the vapor-phase reaction. The micelles in dioxane/water mixture were attached to the surface. The micelle surface was studied by SEM and AFM. The size and polydispersity of the micelles was not altered during the attachment process and the efficiency of the coupling process was high. The surface coverage tended to increase with carbodiimide concentration, but high concentration enhanced aggregation of the micelles. The average diameter of micelle obtained from the SEM pictures was very similar to the hydrodynamic diameter obtained from QELS.

This research suggests several future studies. The kinetics of the micelle attachment process can be studied since a technique to achieve a uniform surface was developed. Chromophores may be reacted with the micelle surface to produce samples appropriate for near-field scanning optical microscopy (NSOM). NSOM imaging could reveal if fluorescence tagging is uniform on the surface.

ACKNOWLEDGMENTS

This research has been supported by the United States Army Research Office (DAAH04-95-1-0127) and the Robert A. Welch Foundation (grant F-356).

REFERENCES

1. (a) Z. Tuzar, P. Kratochvíl, in *Surface and Colloid Science*, E. Matijevic, Ed., Vol. **15**, p.1. Plenum Press, New York 1993; (b) G. Riess, G. Hurtrez, P. Bahadur, In *Encyclopedia of Polymer Science and Engineering*, 2nd ed.; (c) H.F. Mark, N.M. Bikales, C.G. Overberger, G. Menges, Eds., Vol. 2, pp. 324-436. Wiley, New York 1985.
2. (a) M. Tian, C. Arca, Z. Tuzar, S.E. Webber, P. Munk, *J. Polym. Sci.* **33**, 1713 (1995); (b) F. Gadell, W. J. Koros, R.S. Schechter, *Macromolecules* **28**, 4883 (1995); (c) A.V. Kabanov, I.R. Naranova, I.V. Astafieva, E.V. Batrakova, V.A. Alakhov, *Macromolecules* **28**, 2303 (1995); (d) L. Zhang, A. Eisenberg, *Science* **268**, 1728 (1995).
3. (a) M.E. Morrison, Y. Teng, K. Procházka, P. Munk, S.E. Webber, work in progress; (b) E. Arca, M. Tian, S.E. Webber, P. Munk, *Int. J. Polym. Charact.* **2**, 31 (1995); (c) A. Harada, K. Kataoka, *Macromolecules* **28**, 5294 (1995); (d) C. Scholz, M. Iijima, Y. Nagasaki, K. Kataoka, *Macromolecules* **28**, 7295 (1995).
4. (a) P.N. Hurter, T.A. Hatton, *Langmuir*. **8**, 1291 (1992); (b) R. Nagarajan, M. Barry, E. Ruckenstein, *Langmuir* **2**, 210 (1986).
5. (a) T. Cao, P. Munk, C. Ramireddy, Z. Tuzar, S.E. Webber, *Macromolecules* **24**, 6300 (1991); (b) D. Kiserow, K. Procházka, C. Ramireddy, Z. Tuzar, P. Munk, S.E. Webber, *Macromolecules* **25**, 461 (1992); (c) C. Ramireddy, Z. Tuzar, K. Procházka, S.E. Webber, P. Munk, *Macromolecules* **25**, 2541 (1992); (d) M. Tian, A. Qin, C. Ramireddy, S.E. Webber, P. Munk, Z. Tuzar, K. Procházka, *Langmuir* **9**, 1741 (1993); (e) A. Qin, M. Tian, C. Ramireddy, S.E. Webber, P. Munk, Z. Tuzar, *Macromolecules* **27**, 120 (1994); (f) K. Procházka, D. Kiserow, C. Ramireddy, Z. Tuzar, P. Munk, S.E. Webber, *Macromolecules* **25**, 454 (1992); (g) J. Chan, S. Fox, D. Kiserow, C. Ramireddy, P. Munk, S. E. Webber, *Macromolecules* **26**, 7016 (1993); (h) T. Martin, S.E. Webber, *Macromolecules* **28**, 8845 (1995).
6. (a) S. Patel, M. Tirrell, *Annu. Rev. Phys. Chem.* **40**, 597 (1989); (b) F.E. Runge, M.S. Kent, H. Yu, *Langmuir* **10**, 1962 (1994).
7. T. Cao, W. Yin, J.L. Amstrong, S.E. Webber, *Langmuir* **10**, 1841 (1994).
8. (a) J. Chan, Ph.D. Thesis, The University of Texas at Austin, 1994; (b) J. Zhu, R.B. Lenox, A. Eisenberg, *J. Phys. Chem.* **96**, 4727 (1992).
9. M.A. Karymov, K. Procházka, J.M. Mendenhall, T.J. Martin, P. Munk, S.E. Webber, *Langmuir* **12**, 4748 (1996).
10. T.J. Martin, Ph.D. Thesis, The University of Texas at Austin, 1996.
11. Z. Tuzar, private communication.
12. A.W. Adamson, *Physical Chemistry of Surfaces*, 5th ed., p. 387-388. Wiley, New York 1990.

Behavior of Polymers at Surfaces and Interfaces as Observed in Simulated Systems

Tadeusz Pakula

Max-Planck-Institut für Polymerforschung, Postfach 3148, 55021 Mainz, Germany

Abstract: Cooperative motion algorithm (CMA) is used to simulate polymer chains in three types of dense systems reflecting special cases of polymer behavior at surfaces and interfaces: polymer brushes with variable grafting density in the range 0 - 1, both in a neutral solvent and in a polymer melt, layers of end-functionalized polymers between parallel end-adsorbing walls, and copolymers of various distributions of comonomer units (random, block and gradient copolymers) at interfaces with noncompatible polymers.

INTRODUCTION

Behavior of polymers at surfaces and interfaces has been a subject of intensive theoretical, experimental and computer simulation studies for last few years (Refs.1-5). The extraordinary interest in such problems is caused by their relevance to many technologically important systems. The behavior of chains adsorbed or grafted at surfaces or the behavior of chains trapped at interfaces in multicomponent systems are examples of problems associated with this group. In this paper, we use computer simulation to study (1) the behavior of polymer brushes consisting of linear chains grafted at a planar surface, (2) the behavior of linear end-functionalized chains in thin layers, and (3) the behavior of various copolymers acting as compatibilizers at the interface of two noncompatible polymers.

Brushes are considered as model systems and various theoretical treatments have been applied to predict their properties as a function of "hair" densities (Refs. 6-10). We consider here systems of linear chains grafted with one end to a planar surface, with grafting densities ranging between 0 and 1. Equilibrium structures of such brushes are studied in two media: (1) in a neutral solvent and (2) in contact with melts of polymers of various chain lengths. The results of simulation are compared with theoretical predictions.

Polymer layers are often used to bond solid surfaces. Properties of such bonding are dependent not only on the strength of adhesive interaction of the polymer with the solid surface but also on the structure within the polymer layer. A special case of polymers with functionalized ends, improving adhesion to the solid surface, is analyzed. Fraction of chains bridging directly parallel solid surfaces is determined as a function of the layer thickness and interaction strength of the functionalized ends.

Various copolymers differing by distribution of constituent monomer units along the chain have found versatile applications in technology of polymeric materials. Recently a possibility of synthesis of gradient copolymers has been suggested, i.e. of copolymers in which the composition of constituents would change continuously along the chain in a predefined manner. Properties of such polymers have been analyzed by computer simulation (Ref.11), which has indicated that such polymers can provide a possibility of a continuous shift of their dynamic and thermodynamic properties along the temperature scale by changes in composition gradients. Gradient copolymers have now been synthesized (Refs.12,13). Copolymers have often been considered as blend additives diminishing incompatibility at a polymer/polymer interface. Both block (Ref.14) and random (Ref.15) copolymers have been used in

 CCC 1022-1360/99/$ 17.50+.50/0

such applications, each of them affecting the interfacial properties in different way. Here, gradient copolymers are considered as potential compatibilizers and their behavior at interfaces in comparison with block and random copolymers is analyzed.

SIMULATION METHOD

The simulation method used is the Cooperative Motion Algorithm (CMA), described elsewhere (Refs.11,16,17). In this type of simulation, ensembles of beads in a lattice are connected by nonbreakable bonds to objects representing linear macromolecules. Face-centered cubic (fcc) lattice is used with all lattice sites occupied in order to represent dense systems like polymer melts. The systems are considered under the excluded volume condition, which means that each lattice site can only be occupied by a single molecular element (bead). Solvent is represented by single non-bonded beads. In such systems, strictly cooperative dynamics is used consisting in rearrangements satisfying local continuity of the simulated system (no empty lattice sites are generated). This is realized by local motions consisting of displacements of a certain number of molecular elements along closed loops with each element replacing one of its neighbors so that the sum of displacements of elements taking part in the rearrangement is zero (the continuity condition). During such rearrangements, the models of macromolecules undergo conformational transformations preserving, however, their identities given by the number and sequences of elements in the polymer.

In the case of copolymers, two types of monomer units A and B are considered. The two types of units are partially compatible, which is characterized by direct interaction parameters ε_{ij}. In order to generate equilibrium states, a dense system of chains is subjected to motion at a given temperature. Moving a chain element alters the local energy because the monomer units contact new neighbors. An attempt to move a single monomer unit is assumed as one Monte Carlo step and the probability of the movement is related to the interaction energy of the monomer unit in the attempted position. Under such conditions, at low temperatures, different types of monomeric units tend to separate from each other in order to reduce the number of A-B contacts and, consequently, to reduce the energy.

In order to represent polymer layers, such systems are considered between nonpenetrable parallel walls (Refs.5,18). In the case of brushes, grafted ends are assumed to be mobile at the grafting wall. In order to represent the interaction of chain ends with the surfaces, it is assumed that the states at the adsorbing wall and away from it can differ in potential energy by $\Delta\varepsilon$. Consequently, for an attractive interaction, movements involving separation of the functionalized end from the wall have been performed with a reduced probability $p = \exp(-\Delta\varepsilon)$, which led to the adsorption of functionalized ends on the walls for $\Delta\varepsilon > 0$.

RESULTS

Polymer Brushes

Polymer brushes consisting of linear chains (N=80) anchored to a planar surface by one end have been simulated. Various grafting densities and two media surrounding the brushes have been considered. Structures of brushes are characterized by distributions of concentrations of grafted chains and distributions of free chain ends of grafted chains along the distance from the grafting wall. The results for brushes in a neutral solvent are shown in Fig. 1. Strong extension of chains with increasing grafting density is observed. At all grafting densities, brushes are penetrated by the solvent.

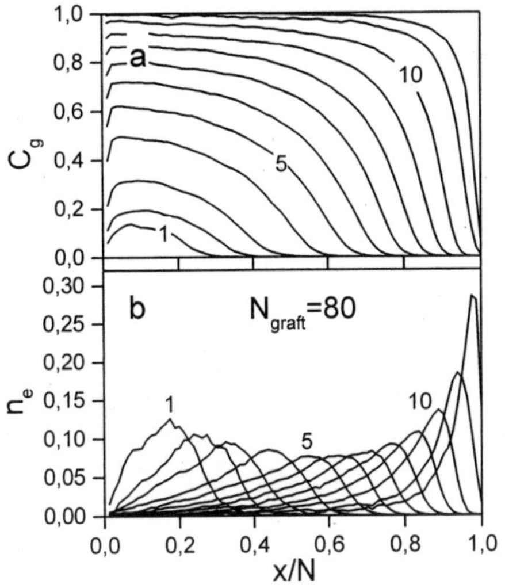

Fig. 1.
Distribution of (a) concentration and (b) free chain ends of brushes with various grafting densities in a neutral solvent along the distance (x) from the grafting wall (normalized by the extended chain length, N).

Grafting densities:

1: 0.025	2: 0.05
3: 0.1	4: 0.2
5: 0.3	6: 0.4
7: 0.5	8: 0.6
9: 0.7	10: 0.8
11: 0.9	12: 0.95

The situation changes when the brush is in contact with a melt of linear chains of the same length as that of grafted chains. The results are shown in Fig. 2.

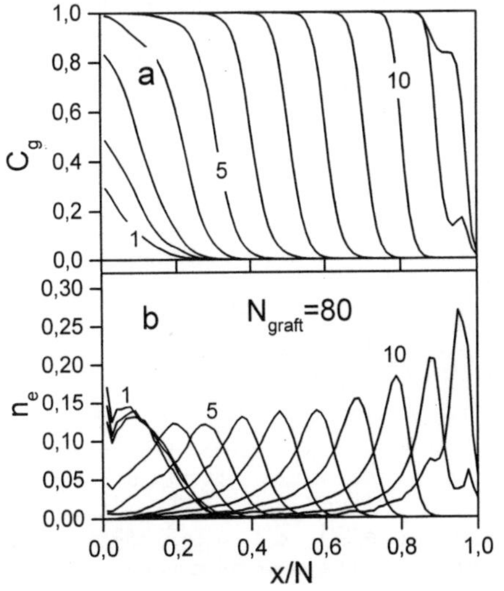

Fig. 2.
Distribution of (a) concentration and (b) free chain ends of brushes with various grafting densities in a polymer melt along the distance (x) from the grafting wall (normalized by the extended chain length, N). For grafting densities, see Fig.1.

Brushes being in contact with the melt already at relatively low grafting density form a dense surface layer not penetrable for other polymer chains having lengths comparable with the lengths of brush hairs. Only shorter chains can partially penetrate such brushes. The results concerning this effect will be presented elsewhere.

In order to characterize dependences of brush thickness on the grafting density, two measures of thickness have been considered: (1) that based on the concentration profile along the distance from the grafting wall (x)

$$t_c = 2\frac{\int xC_g(x)dx}{\int C_g(x)dx} \tag{1}$$

and (2) that based on the distribution of free ends of grafted chains

$$t_e = \frac{\int xn_e(x)dx}{\int n_e(x)dx} \tag{2}$$

Dependences of the brush thickness on the grafting density are compared with various theoretical predictions in Fig. 3. In the case of wet brushes, the self-consistent field theory (Ref.9) overestimates the brush thickness at higher coverage densities. This effect is corrected in another consideration taking into account a finite extensibility of chains (Ref.8). The modified version of the Alexander analysis (Refs.6,10) shows a good agreement with simulation in the whole coverage range if distributions of free ends are taken as a measure of the layer thickness. For high-coverage densities in dry brushes, the results can be well approximated by the thickness of the dense layer (dotted line). Deviations from such dependence are observed at low coverage densities where the concentration profiles reflect dimensions of isolated grafted chains. Distortions in concentration profiles at coverage densities close to unity, observed mainly for dry brushes, are caused by limited sizes of model systems and problems of free chains to arrange two-dimensionally between the brush and the neutral surface.

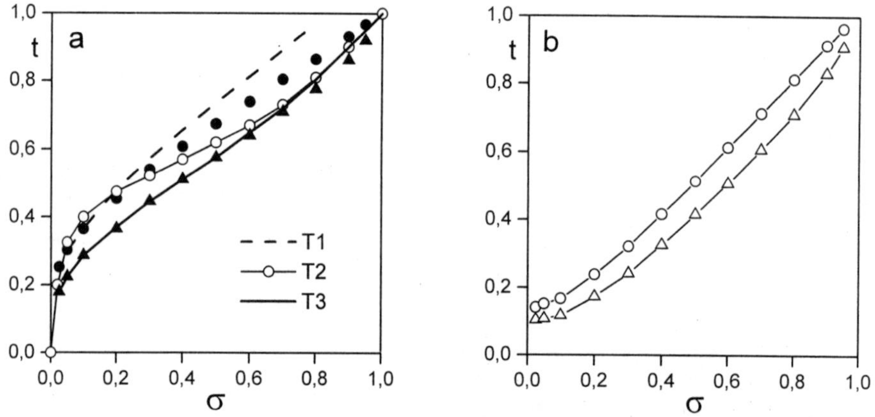

Fig. 3. A comparison of dependences of brush thickness (●,○ t_c; ▲,Δ t_e) on grafting densities determined (a) in a neutral solvent and (b) in a melt, with various theoretical predictions T1 (Ref.9), T2 (Ref.8) and T3 (Ref.10).

End-Functionalized Polymers between Parallel Surfaces

Dense systems of linear chains (N=90) with both ends interacting attractively with two parallel surfaces, thus confining the systems to layers, are considered. The influence of the interaction strength and layer thickness on the structure is analyzed.. Both these parameters considerably influence layer structures which are characterized by distributions of centers of gravity of chains and chain ends as well as by concentration profiles of chains forming loops (both ends adsorbed on the same wall), bridges (ends adsorbed on opposite walls), hairs (one end adsorbed) or remaining free between the walls. The interaction strength is changed in a broad range, neutral walls and nonreversible adsorption of chain ends being the extremes. Figure 4 shows examples of such distributions for various interaction energies at a constant layer thickness.

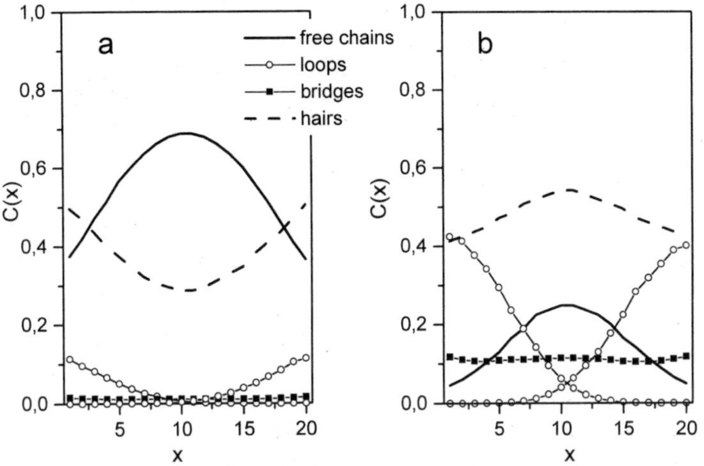

Fig. 4. Comparison of concentration profiles of loops, bridges, hairs and free chains for two energies of interaction between chain ends and walls (d=20): (a) E=1, (b) E=3. The distance from the wall (x) is expressed in units of the lattice constant.

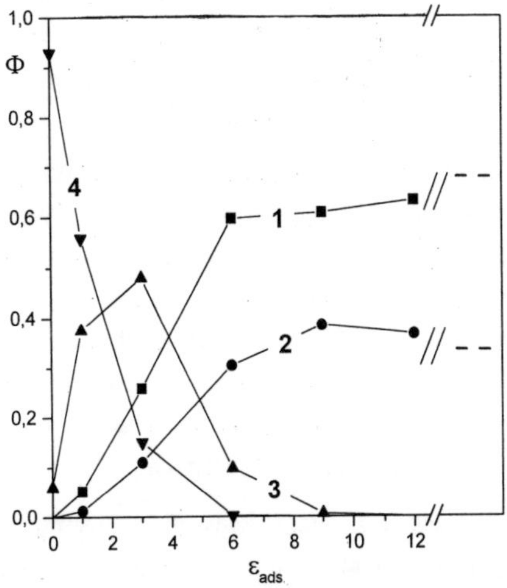

Fig. 5. Fractions of loops (1), bridges (2), hairs (3) and free chains (4) as functions of interaction strength between chain ends and walls (N=90, d=20).

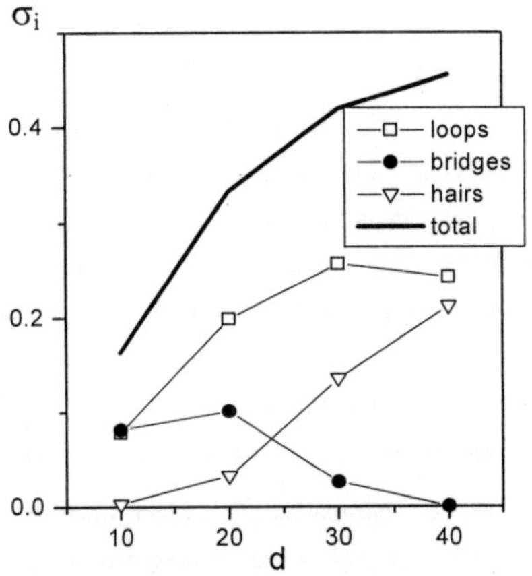

Fig. 6. Coverage of walls by adsorbed chain ends belonging to loops, bridges and hairs as a function of the distance between walls; total coverage is shown by the thick solid line (N=90, E=6).

It can be seen that an increasing adsorption of chains with increasing interaction strength considerably changes the distribution of chains with different roles in the layer. Changes in fractions of chains in different states related to end adsorption at the walls are presented in Fig. 5 as a function of the interaction energy. For a given thickness of the layer, comparable with chain sizes, all chain ends are able to be adsorbed on the walls at a sufficiently high interaction energy. Nevertheless, in such states most chains form loops and only about 35 % of chains contribute to bonding the walls by bridges. It is interesting to note that for irreversibly adsorbing chains, the fraction of bridges is even smaller. Proportions of various chain states are strongly dependent on the layer thickness. This effect is illustrated in Fig. 6. A maximum density of chains bridging the walls per unit wall surface is observed at a layer thickness corresponding to a quarter of the extended chain length. This shows that bonding of surfaces by end-functionalized polymers can be effective only at extremely small layer thickness.

Copolymers as Compatibilizers

Poor adhesion between incompatible polymers is caused by a reduction in cohesive forces between molecular segments of different type but also by specific molecular configurations which can be expected if regions of one polymer are not penetrable by chains of the other. Changes in polymer chain configuration near the interface between incompatible polymers with respect to a virtual interface in a homopolymer are illustrated in Fig. 7 by means of distributions of various molecular parameters along an axis perpendicular to the interface. Concentration profiles (Fig. 7a), distributions of gravity centers of chains (Fig. 7b) and chain ends (Fig. 7d), distributions of chain orientations (Fig. 7c) as well as distributions of local A-B interactions (Fig. 7e) are considered.

The virtual interface in a homopolymer is obtained by considering chains having their gravity centers on different sides of a plane, chosen arbitrarily in a system, as different components. The interface between incompatible polymers is simulated by equilibrating a system of two initially separated polymer layers, brought to a contact at a temperature far below the critical point of the corresponding blend. In this case, it has been proved that identical equilibrium states are obtained if the interactions between chains of type A and B distinguished in a homopolymer by the virtual interface are switched.

The results indicate considerable changes in the molecular structure which the incompatibility of polymers can cause at the interface. The most drastic changes observed consist in removing of gravity centers of chains from the center of the interface, in an increase in chain end concentration in the center of the interface and in a considerable reorientation of chain axes in the directions parallel to the interface. Considerably smaller effects concern differences in concentration profiles. Moreover, it is interesting to note that the range of the mentioned structural effects is remarkably broader than the interface thickness if considered as a width of the concentration change. All the above effects lead to strong heterogeneities in the system and all can contribute to weakness of the interface.

An effective compatibilizer should maximally reduce these heterogeneities by smoothing the distributions of chain ends, chain gravity centers and chain orientations to the levels characteristic of homopolymers. Figure 7 shows how effectively these heterogeneities can be suppressed by various types of copolymers added to the system (again equilibrium states are considered). It can be seen that different types of copolymers involve different changes in the interface structure. There is generally a relatively weak effect on concentration distributions. With a diblock copolymer, the concentration profile

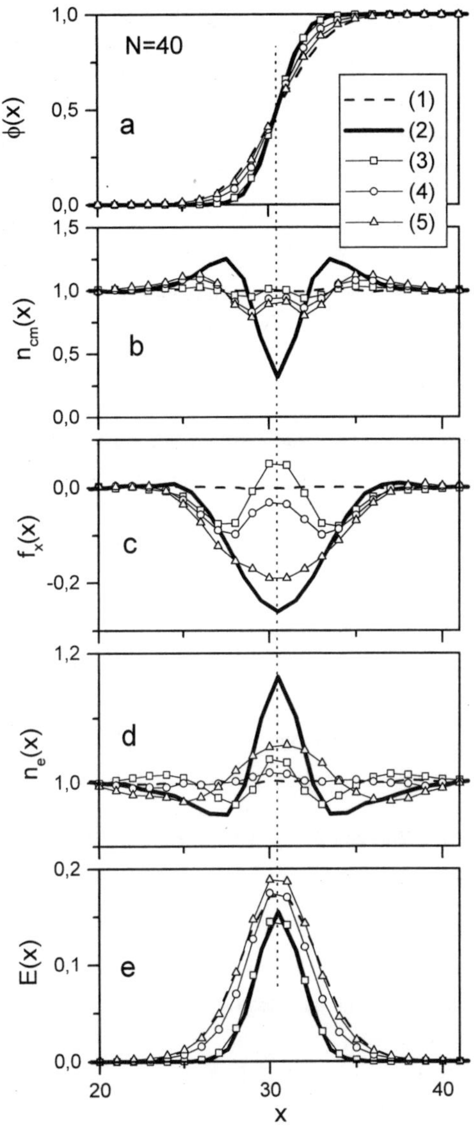

Fig. 7.

Distributions of (a) concentration, (b) gravity centers of polymer chains, (c) orientation factor of chain end-to-end vectors with respect to the normal to the interface, (d) chain ends, and (e) A-B contacts along a direction perpendicular to the interface (x - distance, in units of the lattice constant along this direction).

Interfaces: (1) a virtual interface in a homopolymer, (2) A-B interface of strongly incompatible polymers and A-B interfaces modified with (3) diblock, (4) gradient, and (5) random copolymers.

and, consequently, the profile of active A-B interactions remains almost unchanged with respect to the nonmodified interface. Both the random and gradient copolymers make the concentration profiles distinctly broader and very close to the concentration profile of the virtual interface in a homopolymer. All copolymers strongly suppress nonuniformities in the distribution of chain centers of gravity but the strongest effect is observed for the diblock copolymer. Considerable reduction in chain orientation preferences is observed only for block and gradient copolymers. Random copolymers do not reduce orientations of polymers in the vicinity of the interface and they themselves assume orientations parallel to the interface. The inhomogeneity of chain ends distribution is most effectively reduced by the gradient copolymer for which almost uniform distribution, like in a homopolymer, is obtained.

CONCLUSIONS

The structure of polymer brushes is considerably influenced by a surrounding medium. Neutral solvents can penetrate the brush at all hair densities but linear chains of lengths comparable with the lengths of hairs remain excluded from the brush. Various theoretical descriptions of brush thickness as a function of hair density describe only approximately the behavior of simulated systems.

End-functionalized polymers between end-adsorbing walls can effectively bond the walls only if the layer thickness is considerably smaller than the lengths of extended polymer chains.

Simulations of effects which can be expected at interfaces of incompatible polymers when modified by copolymers with various distributions of monomer units suggest that gradient copolymers could be considered as effective compatibilizers combining advantageous properties of diblock and random copolymers.

Simulations by cooperative rearrangements (CMA) allow to effectively study various phenomena related to the behavior of dense polymer systems at surfaces and to the behavior of chains at interfaces in polymer melts.

ACKNOWLEDGEMENT

The author thanks to E. B. Zhulina, A. Halperin and K. Matyjaszewski for stimulating discussions and cooperation.

REFERENCES

(1) A. Halperin, M. Tirell, T.P. Lodge, Adv. Polym. Sci. **100**,31(1992)

(2) G.S. Grest, M. Nurat, in "Monte Carlo and Molecular Dynamics Simulation in Polymer Science", Ed. K. Binder, Oxford University Press, New York 1995.

(3) S.H. Anastasiadis, T.P. Russel, S.R. Satija, J.F. Majkrzak, J. Chem. Phys. **92**,5677(1990)

(4) M. Stamm, in "Physics of Polymer Surfaces and Interfaces", Ed. J.C. Sanchez, Butteworth-Heinemann, Boston (1992)

(5) E. B. Zhulina, T. Pakula, Macromolecules **25**,754(1992)

(6) S. Alexander, J. Phys. (Paris) **38**,977(1977)

(7) E.B. Zhulina, O.V. Borisov, V.A. Priamitsyn, J. Colloid Interface Sci. **137**,495(1990)

(8) D.F.K. Shim, M.E. Cates, J. Phys. (Paris) **50**,3535(1989)

(9) E.B. Zhulina, O.V. Borisov, L. Brombacher, Macromolecules **24**,140(1991)

(10) P.Y. Lai, A. Halperin, Macromolecules **24**,4981(1991)

(11) T. Pakula, K. Matyjaszewski, Macromol. Theory Simul. **5**,987(1996)

(12) J.S. Wang, K. Matyjaszewski, J. Am. Chem. Soc. **117**,5614(1995)

(13) D. Greszta, K. Matyjaszewski, Polym. Prepr. **37**(1),569(1996)

(14) R. Fayt, R. Jerome, Ph. Teyssie, J. Polym. Sci. Polym. Phys. Ed. **27**,775(1989)

(15) C.A. Dai, B.J. Dair, K.H. Dai, C.K. Ober, E.J. Kramer, C.Y. Hui, L.W. Jelinski, Phys. Rev. Lett. **73**,2472(1994)

(16) A. Gauger, A. Weyersberg, T. Pakula, Macromol. Chem. Theory Simul. **2**,531(1993)

(17) T. Pakula, Recent Res. Dev. Polym. Sci. **1**,101(1996)

(18) T. Pakula, E.B. Zhulina, J. Chem. Phys. **95**,4691(1991)

Macromol. Symp. ***139,*** *57–65 (1999)*

Polysoaps and Biopolymers: The Effect of Secondary Structure on the Extension Elasticity

Oleg V. Borisov[1] and Avraham Halperin*[2]

[1]BASF AG ZOI/ZC-C13, 67056 Ludwigshafen, Germany
[2]DRFMC-SI3M CEA-Grenoble 38054 Grenoble Cedex 9, France

Abstract: Extension of DNA and the protein titin reveals a plateau in the force *vs.* end-to-end distance diagram. A similar feature is predicted for the stretching elasticity of long polysoaps. These are linear hydrophilic chains that incorporate, at intervals, covalently bound amphiphilic monomers. In water, the amphiphilic monomers self-assemble into intrachain micelles thus endowing the chain with a secondary structure. The similarity in the force laws is traceable to the equilibration of the secondary structure. For polysoaps forming a linear string of micelles, the plateau is associated with a coexistence of weakly perturbed micelles and fully dissociated amphiphiles. Similar scenarios can be invoked to explain the behaviour of DNA and titin. The resulting functional form seems to depend only weakly on the precise nature of the system.

Introduction

Recent experimental developments, involving micropipets, laser twizzers and antibodies, enable the measurement of the force laws characterising the extension of certain biopolymers. So far, these studies have been conducted on DNA [1] and the muscle protein titin [2]. The plots of the force, f, versus end-to-end distance, R, reveal a novel feature. An initial regime, consistent with the behaviour of a worm-like chain, is followed by a weak plateau. In the weak plateau, the increase in f with R is much weaker than the $f \sim R$ typical of Gaussian elasticity. A similar force was recently predicted [3, 4, 5] for a family of synthetic polymer, polysoaps [6]. Polysoaps consist of a flexible hydrophilic backbone incorporating, at intervals, covalently bound amphiphilic monomers. As we will discuss, the similarity in the force laws is not spurious. It arises because polysoaps [7, 8], like DNA and proteins, undergo an intramolecular self-assembly leading to a hierarchy of self-organisation. In the following, we will briefly discuss the similarities between the configurations of polysoaps and of biopolymers. We then present a brief discussion of the plateau regime [3, 4].

Borrowing from the protein terminology, we may refer to the monomer sequence as *primary* structure. In water, the amphiphilic monomers self-assemble into intrachain micelles. These comprise of a dense core formed by the hydrophobic tails of the amphiphiles. The polar head groups straddle the core-water boundary. The structure of this inner

 CCC 1022-1360/99/$ 17.50+.50/0

part of the intrachain micelle is similar to the structure of micelles formed by monomeric amphiphiles [9]. The dense inner core is surrounded by a semidilute corona of swollen loops formed by the spacer chains upon aggregation. In the following, we limit the discussion to spherical intrachain micelles. In this case the structure of the corona is reminiscent of star-like polymers [10]. The intrachain micelles may be considered a *secondary* structure. When the number of amphiphiles in the chain, m, exceeds the equilibrium aggregation number of the intrachain micelles, p_{eq}, the polysoaps self-assemble into a string of micelles. In the case of long polysoaps, $m \gg p_{\text{eq}}$, the string may adopt numerous configurations. We refer to the configurations adopted by the string of micelles as *ternary* structure. The simplest configurations to envision are: a linear string, a randomly branched string and a spherical globule of closely packed micelles. This last configuration arises as a consequence of bridging attraction between different micelles due to the exchange of amphiphiles. In polysoaps, this gives rise to spacer chains that bridge different micelles. The globular state is predicted to be the most stable [8]. However, different configurations may occur as long-lived metastable states. This feature arises because configurational rearrangement is an activation process. It involves repartitioning of amphiphiles between different micelles. In turn, this involves, as an intermediate step, the expulsion of amphiphiles from existing micelles into the surrounding aqueous medium. As a result, the free energy surface characterising the configurational phase space of polysoaps is rugged. This feature has been occasionally used to define "complex systems" as opposed to simple systems whose free energy surface exhibits a single minimum [11]. A ruggedness of the free energy surface is also a characteristic of proteins. However, in this case, the ruggedness is due to frustration, a feature which is absent in simple polysoaps comprising a single type of amphiphilic monomers.

Proteins, like polysoaps, exhibit a hierarchy of intrachain self-organisation characterized by a rugged free energy surface. Before we discuss some more similarities, it is important to stress the limitations of this correspondence. Biopolymers, proteins and DNA incorporate a larger number of monomer units and a much richer repertoire of interactions. In this context, directed interactions, such as hydrogen bonds are of special importance. As a result, biopolymers can assume unique structures. In folded proteins, the positions of most atoms are uniquely specified. This detailed structure, and the associated functions, have no counterpart in polysoaps. However, other similarities do exist on a more "coarse-grained", large-scale level. Both proteins and polysoaps [7, 8] unfold, denature when treated with free surfactants. Polysoaps are predicted to undergo both cold and hot denaturation [12] as is the case for proteins, i.e. the self-organisation disappears both at high and low temperatures. The detailed features of the process for polysoaps and for proteins are different. The similarities are traceable to the key role of hydrophobic interactions in driving the self-assembly of the two types of macromolecules. Another common feature concerns the gelation behaviour. Polysoaps, like proteins, form physical gels. These involve secondary structures incorporating monomer units belonging to different polymers. This brings us to the similarity, noted earlier, in the extension elasticities. In the following, we will discuss the plateau regime expected when a *linear* string of spherical micelles is stretched.

The unperturbed polysoap

Our discussion focuses on polysoaps forming spherical, star-like micelles whose size is dominated by the corona. In a θ solvent, the size of such micelles, as given by a Flory-type theory [10], is $r_{\text{micelle}} \approx p^{1/4}n^{1/2}b$ where b is a characteristic monomer size. Modification of simple phenomenological model [9] yields the free energy per amphiphile in a micelle of aggregation number p, $\epsilon_p kT$ [7]

$$\epsilon_p \approx \gamma a_o(u^{-1/3} + u^{1/3} + \kappa u^{1/2}) - \delta \tag{1}$$

Here γkT is the surface tension of the water-core interface, a_o is the optimum area per head group in a micelle formed by non-polymerized amphiphiles and $-\delta kT$ is the transfer free energy of a hydrophobic tail from water into the micellar core. For spherical micelles, p is related to the volume of the hydrophobic tail, v, and to the area per head group, a, as $p \approx v^2/a^3$ and thus $p_o \approx v^2/a_o^3$. u is a dimensionless variable $u = p/p_o = (a_o/a)^3$. The $u^{-1/3}$ term allows for the interfacial free energy per head group and thus favours micellar growth. The repulsion between the head groups gives rise to the $u^{1/3}$ term. The third, $u^{1/2}$, term arises because of the repulsive interactions between the coronal loops. In a θ solvent, the free energy per loop is $F_{\text{corona}}/kT \approx p^{1/2}$ [10]. These last two terms reflect free energy penalties that oppose micellar growth. The relative importance of the penalty terms is determined by $\kappa \approx p_o^{1/2}/\gamma a_o$, the ratio of the coronal penalty and the head group repulsion at $a \approx a_o$. For brevity, we focus on the limit of $\kappa \ll 1$ when $p \approx p_o$ and $u \approx 1$. For $p \approx p_o$, $\epsilon_{p_o} \approx 2\gamma a_o - \delta$ and the critical micelle concentration X_{cmc}, is $\epsilon_{p_o} \approx \ln X_{\text{cmc}}$. It is important to note that a θ solvent for the backbone is, in effect, a good solvent for the string of micelles. The θ temperature for star polymers is lower than the θ temperature of the corresponding linear chain [13]. Accordingly, the micellar coronas do not interpenetrate significantly. A string of intrachain micelles is thus self-avoiding and its span is the Flory radius $R_{\text{F}}(p) \approx (m/p)^{3/5} r_{\text{micelle}}(p) \approx R_{\text{F}} u^{-7/20}$. Here $R_{\text{F}} \approx (m/p_o)^{3/5} r_{\text{micelle}}(p_o) \approx m^{3/5}n^{1/2}p_o^{-7/20}b$ is the Flory radius of a string of spherical intrachain micelles when $p \approx p_o$.

Our discussion is based on the assumption that all polymerized amphiphiles in the unperturbed chain are micellized. The validity of this assumption depends on m. It is untenable once m becomes larger than a certain m_u to be specified. For $m \approx m_u$, an average of one micelle per chain is fully dissociated. The underlying physics is as follows. The dissociation is favoured by the mixing entropy of the one-dimensional mixture of micelles and dissociated amphiphiles, S_{mix}. In turn, the importance of this mixing entropy grows with m. To specify m_u, it is first necessary to obtain an explicit expression for S_{mix} as a function of the fraction of associated amphiphiles, ψ. This involves inscription of the micelles and the dissociated amphiphiles on a one-dimensional lattice comprising M_ψ sites. The distinctive feature of this system is that M_ψ varies with ψ. Since the number of micelles is $\psi m/p_o$ and the number of dissociated amphiphiles is $m(1-\psi)$, $M_\psi \approx m(1-\psi)+\psi m/p_o$. The associated $S_{\text{mix}}(\psi)$ is the ideal mixing entropy $S_{\text{mix}}/k = -M_\psi[Y \ln Y + (1-Y)\ln(1-Y)]$ where Y is the fraction of sites occupied by micelles, $Y \approx \psi m/p_o M_\psi$. This leads to

$$S_{\text{mix}}(\psi)/k \approx \frac{\psi m}{p_o}\ln[1+\frac{p_o(1-\psi)}{\psi}] + m(1-\psi)\ln[1+\frac{\psi}{p_o(1-\psi)}] \tag{2}$$

For future reference, it is useful to provide the expression for $S'_{\text{mix}} \equiv \mathrm{d}S_{\text{mix}}/\mathrm{d}\psi$

$$S'_{\text{mix}}(\psi)/k \approx \frac{m}{p_o}\ln[1+\frac{p_o(1-\psi)}{\psi}] - m\ln[1+\frac{\psi}{p_o(1-\psi)}] \tag{3}$$

The free energy of a partially dissociated chain of micelles is $F_{\text{chain}}/kT = m\psi\epsilon_{p_o} - S_{\text{mix}}(\psi)/k$ describing an ideal one-dimensional mixture of dissociated amphiphiles and intrachain micelles. To consider the onset of micelle dissociation, it is helpful to consider F_{chain} for $\psi \approx 1$ when $S_{\text{mix}}/k \approx -m(1-\psi)\ln p_o(1-\psi)$. It is convenient to write it as

$$F_{\text{chain}} \approx F_o + m|\epsilon_{p_o}|(1-\psi) + m(1-\psi)\ln p_o(1-\psi) \tag{4}$$

where $|\epsilon_{p_o}|$ is the free energy cost per polymerized amphiphiles for the dissociation of a single micelle. The form of F_{chain} is similar to that of the free energy used to demonstrate the impossibility of the coexistence of two semi-infinite phases in a one-dimensional system experiencing short-range interactions [14]. In the present case, we can identify the micellized and dissociated states with the two phases. The assumption of a fully micellized chain amounts to assuming a single, defect-free, one-dimensional phase of finite size. The equilibrium condition $\partial F_{\text{chain}}/\partial\psi = 0$ leads to $p_o\alpha(1-\psi) \approx \exp(-|\epsilon_{p_o}|)$. It is helpful to recast this result in terms of the average number of dissociated micelles per chain, n_{d}. This is related to ψ as $\psi \approx 1 - n_{\text{d}}p_o/m$ leading to $n_{\text{d}} \approx p_o^{-2} m \exp(-|\epsilon_{p_o}|)$. The condition $n_{\text{d}} = 1$ defines m_u

$$m_u \approx p_o^2 \exp(|\epsilon_{p_o}|) \tag{5}$$

and thus

$$n_{\text{d}} \approx m/m_u \tag{6}$$

The "single-phase approximation" is thus justified while $m \ll m_u$ or, equivalently, $n_d \ll 1$. This condition is easily fulfilled since $m_u \gg 1$ because $p_o \gg 1$ and $X_{cmc} \ll 1$.

Extension of linear string of micelles

The elasticity of a linear string of micelles is expected to reflect a coupling to its "internal degrees of freedom", i.e., the micellar structure. According to the Le Chatelier principle [14], the equilibration of the micellar structure should lower the tension in the chain [3, 4, 5]. To obtain an explicit force law, it is necessary to consider the statistical mechanics of the model. The simplest possible scenario involves uniform extension. It is assumed that the extension results only in repartitioning of the amphiphiles among a greater number of smaller micelles. The analysis of this scenario leads to an fR diagram exhibiting, for intermediate extensions, a van der Waals loop. This suggests that, for this range, weakly perturbed micelles coexist with fully dissociated amphiphiles. The weak and strong deformation behaviour is unmodified. The stretching free energy of the string, F_{el} may assume three forms. For weak deformations, when linear response is expected, $F_{el} \approx [R/R_F(p)]^2 \approx (R/R_F)^2 u^{7/10}$. This initial Gaussian F_{el} is replaced by the Pincus free energy [15], $F_{el} \approx [R/R_F(p)]^{5/2} \approx (R/R_F)^{5/2} u^{7/8}$ when $R \gg R_F(p)$. Eventually, when R exceeds the length of the fully extended string of micelles, $R_{max} \approx (m/p) r_{micelle}(p)$, a third regime comes into play. In it, the bridges between the micelles are stretched. Since the backbone experiences a θ solvent, the stretching involves a linear string of m/p Gaussian bridges consisting each of n monomers. Consequently, F_{el} is Gaussian with $F_{el} \approx [R/R_B(p)]^2 \approx (R/R_B)^2 u$ where $R_B \approx (m/p_o)^{1/2} n^{1/2} b$. The relative importance of the elastic penalty is measured by the dimensionless parameter $\tau(R)$. $\tau(R) \approx F_{el}(p_o)/kT\gamma a_o m$ is the ratio of the chain elastic energy and $m\epsilon_p$ for $p \approx p_o$ and a given R. The boundary of the instable region is identified by $\tau(R_*) \approx 1$. It occurs in the stretched bridges regime, for $R_* \approx R_B(m\gamma a_o)^{1/2}$ and $f_* \approx (m\gamma a_o)^{1/2}/R_B$.

When the role of S_{mix} is ignored, the coexistence regime exhibits features reminiscent of a first-order phase transition. In particular $f \sim R^o$. Physical insight concerning this behaviour may be gained by considering two free energies: the free energy of a stretched, fully dissociated chain, F_{dis}, and the free energy of an extended string of unperturbed micelles in the stretched bridges regime, F_B. In the stretched bridges regime, the leading contributions are $F_B/kT \approx m\epsilon_{p_o} + (R/R_B)^2$. The fully dissociated chain is described by $F_{dis}/kT \approx (R/R_o)^2$. The bottom of the free energy curve described by F_B is located at $R_B < R_o$ and is $m|\epsilon_{p_o}|$ below the minimum of F_{dis}. For weak extensions, F_B is lower than F_{dis}. However, $R_B < R_o$ and thus the spring constant of the fully dissociated chain is weaker. Consequently, the two curves cross at R_{co} specified by $F_B \approx F_{dis}$ or $R_{co} \approx$

$R_o[m|\epsilon_{p_o}|/(p_o-1)]^{1/2}$. For $R < R_{co}$ the string of micelles is of lower free energy while for $R > R_{co}$, the lower free energy corresponds to the fully dissociated chain. This cross-over is a rough diagnostic for a first-order phase transition. This rough argument locates the cross-over at $R_{co} \approx (R_U R_L)^{1/2}$. The peak of the "spinodal curve" $R_o < R_* < R_{co}$ corresponds to $(R/R_B)^2 \approx m|\epsilon_{p_o}|$, i.e., when the free energy of the micellar string is equal to the minimal free energy of the dissociated chain. A more detailed analysis proceeds as follows. At coexistence, only a fraction, $0 < \psi < 1$, of the amphiphiles will be incorporated into micelles. Thus, ψm amphiphiles will form weakly perturbed intrachain micelles while the remaining $(1-\psi)m$ surfactants will be dissociated. Since the onset of coexistence is expected to occur at $f_{co} \ll f_*$, we neglect the weak perturbation of the micellar structure and assume $u \approx 1$ or $p \approx p_o$. In a θ solvent, for a given R and T, the free energy of the polysoap is

$$F_{\text{chain}}/kT \approx -\psi m|\epsilon_{p_o}| + R^2/R_B^2[\psi + (1-\psi)p_o] \tag{7}$$

The first term allows for the free energy of the aggregated amphiphiles while the second reflects the Gaussian elasticity of a polysoap with $\psi m/p_o$ intrachain micelles. Within the $S_{\text{mix}} = 0$ approximation, the coexistence is reminiscent of a first-order phase transition and $f \sim R^o$ is expected [3, 4]. It is thus appropriate to analyze the equilibrium behaviour for constant T and f. The corresponding equilibrium condition, as stated in terms of $G_{\text{chain}}(\psi, R) = -\psi m|\epsilon_{p_o}| + F_{\text{el}} - fR$, is $\mathrm{d}G_{\text{chain}} = (\partial G_{\text{chain}}/\partial\psi)\mathrm{d}\psi + (\partial G_{\text{chain}}/\partial R)\mathrm{d}R = 0$. At coexistence, f does not specify a unique R. Rather, $f = f_{co}$ for any $R_L < R < R_U$. The equilibrium condition in this range is thus $\mathrm{d}G_{\text{chain}}/kT = -m|\epsilon_{p_o}|\mathrm{d}\psi - (f_{co}/kT)\mathrm{d}R = 0$ and

$$\mathrm{d}\psi/\mathrm{d}R \approx -f_{co}/kTm|\epsilon_{p_o}| \tag{8}$$

Integration of Eq. 8, together with the condition $\psi(R_U) = 0$ yields

$$\psi \approx (f_{co}/kTm|\epsilon_{p_o}|)(R_U - R).$$

By invoking $\psi(R_L) = 1$, we obtain

$$f_{co}/kT = m|\epsilon_{p_o}|/(R_U - R_L) \tag{9}$$

and

$$\psi \approx (R_U - R)/(R_U - R_L) \tag{10}$$

To obtain the explicit expressions for f_{co}, R_L and R_U, we utilize the following three conditions: $f_{co}/kT \approx R_U/R_o^2 \approx R_L/R_B^2$. These lead to $R_U \approx p_o R_L$, and to

$$R_L \approx [m|\epsilon_{p_o}|/p_o(p_o-1)]^{1/2}R_o;\, R_U \approx [m|\epsilon_{p_o}|p_o/(p_o-1)]^{1/2}R_o \tag{11}$$

and, finally, to

$$f_{co}/kT \approx [m|\epsilon_{p_o}|p_o/(p_o-1)]^{1/2}R_o^{-1} \approx |\epsilon_{p_o}|^{1/2}[p_o/(p_o-1)]^{1/2}(nb^2)^{-1/2} \tag{12}$$

Upon further extension, $R > R_U$, the polysoap is fully unravelled, its free energy is $F_{dis}/kT \approx R^2/R_o^2$ and the corresponding force law is $f/kT \approx R/R_o^2$. Eventually, when the polysoap approaches its fully extended state, its elasticity will be described by the inverse Langevin $\mathcal{L}^*$ form.

In reality, $S_{mix} \neq 0$. This, in turn, introduces modifications of the fR diagram at coexistence. The modified fR diagram is specified by

$$F_{chain}/kT \approx m\psi\epsilon_{p_o} + (R/R_B)^2[\psi(1-p_o)+p_o]^{-1} - S_{mix}/k \tag{13}$$

The elastic term may be viewed as an external field favouring the dissociated "phase". An increase in f, or in the imposed R, always favours a decrease in ψ. On the other hand, S_{mix} always favours intermediate values of ψ. It thus lowers $\tilde{f}_{co}/f_{co}$ in the vicinity of $\psi = 1$ while increasing it in the neighbourhood of $\psi = 0$. Here the tilde is used to distinguish the equilibrium values obtained when allowing for S_{mix} from those obtained previously. When $S_{mix} \neq 0$, $f_{co} \neq \text{const}'$ and we can no longer use the simplified analysis described earlier. Furthermore, when we allow for S_{mix}, it is impossible to solve for $\tilde{f}_{co} = \tilde{f}_{co}(R)$ explicitly. Instead, we can obtain a parametric solution. In particular, we solve for the equilibrium values of $\tilde{R} = \tilde{R}(\psi)$ and of $\tilde{f}_{co} = \tilde{f}_{co}(\psi)$. The condition $\partial F_{chain}/\partial\psi = 0$ yields

$$\tilde{R} = R_B[m|\epsilon_{p_o}| + S'_{mix}/k]^{1/2}[\psi(1-p_o)+p_o]/(p_o-1)^{1/2} \tag{14}$$

Upon substituting this $\tilde{R}$ into Eq. 13, we obtain $F_{chain}(\psi)$ in equilibrium. The tension at coexistence, as a function of ψ, is

$$\tilde{f}_{co} = \mathrm{d}F_{chain}(\psi)/\mathrm{d}R = (\mathrm{d}F_{chain}/\mathrm{d}\psi)(\mathrm{d}R/\mathrm{d}\psi)^{-1}$$

or

$$\tilde{f}_{co} = f_{co}(1 + S'_{mix}/km|\epsilon_{p_o}|)^{1/2} \tag{15}$$

where f_{co} is the tension at coexistence as given by Eq. 12, when S_{mix} is neglected. For $0 \ll \psi \ll 1$, the fR diagram now exhibits a logarithmic correction to $f_{co} \sim R^o$. $\tilde{f}_{co}$ increases weakly with R. S'_{mix} is negligible in the vicinity of $\psi_o \approx 1 - p_o^{-1/2}$ and the corresponding $\tilde{R}(\psi_o) \approx p_o^{1/2}R_L$. Consequently, in this region, it is possible to replace Eq. 14 by Eq. 10. In turn, Eq. 10 yields the lever rule, $\psi/(1-\psi) \approx (R_U - R)/(R - R_L)$. By substituting it into S'_{mix}, it is possible to approximate $\tilde{f}_{co}$ in this region by

$$\tilde{f}_{co} = f_{co}(1 + \ln g(R)/|\epsilon_{p_o}|)^{1/2} \tag{16}$$

where $g(R)$ is a slowly increasing function of R

$$g(R) \approx [1 + \frac{1}{p_o}(\frac{R_U - R}{R - R_L})][1 + p_o(\frac{R - R_L}{R_U - R})]^{1/p_o} \tag{17}$$

These logarithmic corrections become important in the immediate neighbourhood of $\psi = 1$ and $\psi = 0$ where the above approximation is no longer valid. To identify the upper and lower boundaries of the coexistence regime, $\tilde{R}_{\rm U}$ and $\tilde{R}_{\rm L}$, it is necessary to carefully specify the corresponding values of ψ. As was stated earlier, it is assumed that prior to the onset of coexistence, all amphiphiles are assembled into micelles. In the coexistence regime, we further assume that fully dissociated amphiphiles coexist with weakly perturbed micelles. Within this approximation, the lower and the upper boundaries of the coexistence regime correspond, respectively, to chains supporting one dissociated micelle and to chains supporting a single undissociated micelle. In other words, the maximal and minimal values of ψ at coexistence are respectively $\psi_{\rm max} \approx 1 - p_o/m$ and $\psi_{\rm min} \approx p_o/m$. The $S_{\rm mix} = 0$ approximation allows us to consider arbitrarily large m and thus to set $\psi_{\rm max} \approx 1$ and $\psi_{\rm min} \approx 0$. In our case, however, we are limited to $m \ll m_u$ and this is no longer possible. To obtain $\tilde{R}_{\rm L}$ and $\tilde{R}_{\rm U}$, we use $\tilde{f}(\psi_{\rm max})/kT \approx \tilde{R}_{\rm L}/R_{\rm B}^2$ and $\tilde{f}(\psi_{\rm min})/kT \approx \tilde{R}_{\rm U}/p_o R_{\rm B}^2$. Altogether

$$\tilde{R}_{\rm L} \approx R_{\rm L}[1 + \frac{\ln(p_o^2/m)}{|\epsilon_{p_o}|}]^{1/2} \tag{18}$$

$$\tilde{R}_{\rm U} \approx R_{\rm U}[1 + \frac{\ln m}{p_o|\epsilon_{p_o}|}]^{1/2} \tag{19}$$

Note that both $\tilde{f}_{\rm co}(\psi_{\rm max})$ and $\tilde{R}_{\rm L}$ decrease as m increases and vanish when $\ln(p_o^2/m) + |\epsilon_{p_o}| = 0$. In particular, both $\tilde{f}_{\rm co}(\psi_{\rm max})$ and $\tilde{R}_{\rm L}$ vanish as $|\epsilon_{p_o}|^{-1}\ln(m/m_u)$ when $m \to m_u$. This is due to the growing importance of $S_{\rm mix}$ in destabilizing the micelles in the unperturbed chain.

Concluding remarks

Our understanding of the statistical mechanics of polysoaps is at an early stage. The problem is of practical interest because of the numerous applications of polysoaps as viscosity modifiers. A better understanding of their molecular design parameters and their effect on their performance is clearly desirable. In addition, the behaviour of synthetic polymers with secondary structure is of fundamental interest. The study of polysoaps may prove interesting also because of their similarities to biopolymers. The fR diagram discussed above, and its modifications by denaturing agents [5], is an example of possible benefits to the biopolymer community.

References

[1] (a) P. Cluzel, A Lebrun, C. Heller, R. Lavery, J.-L. Viovy, D. Chatenay and F. Caron *Science* **271**, 792 (1996) (b) S. B. Smith, Y. Cui and C. Bustamante *Science* **271**, 795 (1996).

[2] (a) M.S.Z. Kellermayer, S.B. Smith, H.L. Granzier and C. Bustamante *Science* **276**, 1112 (1997) (b) M. Rief, M. Gautel, F. Oesterhelt, J.M. Fernandez and H. Gaub *Science* **276**, 1109 (1997) (c) L. Tskhovrebova, J. Trinick, J.A. Sleep and R.M. Simmons *Nature* **387**, 308 (1997).

[3] O.V. Borisov and A. Halperin *Europhys. Lett.*, **34**, 657 (1996).

[4] O.V. Borisov and A. Halperin *Macromolecules*, **30**, 4432 (1997).

[5] O.V. Borisov and A. Halperin *Phys.Rev.E*, **30**, 812 (1998).

[6] (a) L. Gros, H. Ringsdorf and H. Scupp *Angew. Chem. Int. Ed. Engl.*, **20**, 305 (1981). (b) A. Laschewsky, *Adv. Polym. Sci.*, **124**, 1 (1995). (c) T. E. Hogen-Esch and E. Amis. *Trends Polym. Sci.*, **3**, 98 (1995). (d) E.D. Goddard, and K.P. Ananthapadmanabhan, ed. *Interactions of Surfactants with Polymers and Proteins* CRC Press, Boca Raton, 1993. (e) J.E. Glass, ed. *Hydrophilic Polymers: Performance with Enviromental Acceptance*, American Chemical Society 1996.

[7] O.V. Borisov and A. Halperin *Langmuir* **11**, 2911 (1995).

[8] O.V. Borisov and A. Halperin *Macromolecules* **29**, 2612 (1996).

[9] J.N. Israelachvili, *Intramolecular and Surface Forces*, 2nd ed., Academic Press, London **1991**.

[10] (a) M. Daoud and J.P. Cotton, *J. Phys. (Paris)*, **43**, 531 (1982) (b) E.B.Zhulina *Polym. Sci. USSR*, **26**, 794 (1984).

[11] H. Frauenfelder and P.G. Wolynes, *Phys. Today* **1994** (Feb), 58.

[12] D.E. Leckband, O.V. Borisov and A. Halperin *Macromolecules*, **31**, 2368 (1998).

[13] E.B. Zhulina, O.V. Borisov, T.M. Birshtein, *Polym. Sci. USSR*, **30**, 780 (1988).

[14] L.D. Landau and E.M. Lifshitz, *Statistical Physics* Pergamon Press, Oxford, **1986**.

[15] P. Pincus *Macromolecules* **9**, 386 (1976).

Unusual Adsorption Behavior of Polymers in Layered Structures

Jens-Uwe Sommer* and Alexander Blumen

Universität Freiburg, Theoretische Polymerphysik, Hermann-Herder Str. 3, D–79104 Freiburg, Germany

Abstract: We studied the conformations of polymer chains embedded in regular lamellar matrices such as microphase-separated block copolymer melts or lipid/water systems. In such quasi-one-dimensional systems, the host plays the role of a nearly-periodic external potential. A probe chain can get trapped with the localization length L at isolated defects, when its degree of polymerization N exceeds a cross-over value N_L. Interestingly, the radius of gyration, R_g, of the chain displays a minimum as a function of the defect density, ρ, the minimum being estimated by $\rho_c L \simeq N_L/N$. Using Monte Carlo simulation, we underline the non-monotonic behavior of R_g and calculate Green's function for various defect densities. On the same vein, the mean square displacement of the monomer units at large times shows a pronounced minimum as a function of the defect density, ρ.

Introduction

Microphase-separated block copolymers as well as lipid/water systems can exhibit extended layered structures, patterns which are nearly periodic viewed perpendicular to the interfaces. A long polymer chain embedded in such a host is then subject to a periodic external potential, given that each monomer unit is more attracted to one of the phases than to the other [1, 2]. An alternating AB copolymer whose repeat unit is a short diblock is located preferentially at the interfaces, provided that the host components act selectively on the A and the B species [3, 4, 5]. The interfaces are then adsorbing planes for the repeat units. Common to all of these situations is a one-dimensional (1D) periodic, external potential acting on a tagged chain. This is an interesting problem in theoretical polymer physics since equilibrium statistical properties of ideal chains in 1D potentials can be often obtained exactly. Moreover, as it is well known from the solid state physics [6], disorder effects are most pronounced in 1D. This makes it possible to investigate polymers in disordered environments under extreme conditions and from a new point of view.

Polymers in periodic potentials

We start by displaying the effect of a perfectly periodic potential of the Kronig-Penney form [7] on the chain conformations by using transfer matrix methods. The results can be extended to account for isolated defects of the periodicity. We will show that these defects can localize the chain and that the problem is closely related to the appearance of defect levels in the band structure of solids. Taking into account a finite density of defects, we find that the extension of the chain has a minimum as a function of the defect

 CCC 1022-1360/99/$ 17.50+.50/0

density, ρ. We support our results by Monte Carlo simulations; these also allow insights into the dynamics of chains in 1D disordered media. Our problem is furthermore closely related to the motion of interface lines on a striped surface, as recently considered by Nieuwenhuizen [8].

Here and in the following, we consider an ideal chain of polymerization degree N in an external potential $V(x)$ (given in units of kT). In this case the partition function decouples with respect to the Cartesian x, y and z coordinates. Hence we focus only on the x direction, taken perpendicular to the layers. Green's function $G(x, x'; N)$ of a chain which starts at x' and ends at x obeys the following equation [9]:

$$\frac{\partial}{\partial N} G(x, x'; N) = \frac{l^2}{2} \frac{\partial^2}{\partial x^2} G(x, x'; N) - V(x) G(x, x'; N) \tag{1}$$

with $G(x, x'; 0) = \delta(x, x')$. In Eq. (1), l is the statistical segment length of the chain. For simplicity, we use dimensionless units for all quantities and set $l^2 = 2$ as well as $kT = 1$.

Let us start from an unperturbed periodic potential of the Kronig-Penney form:

$$V(x) = \sum_{\alpha=-\infty}^{\infty} \epsilon \delta(x - \alpha\xi) \tag{2}$$

Depending on the prevailing conditions, ϵ can be either positive or negative. The model given in Eqs. (1) and (2) reproduces the basic features of chains in layered hosts and has the advantage of rendering the mathematics simple. In the following we call $\epsilon < 0$ ($\epsilon > 0$) the attracting (repulsive) case. Now Green's function can be represented as

$$G(x, x', N) = \sum_k e^{-\lambda_k N} \phi_k(x) \bar{\phi}_k(x') \tag{3}$$

where λ_k and ϕ_k are the eigenvalues and eigenfunctions of the equation

$$-\frac{\mathrm{d}^2}{\mathrm{d}x^2} \phi_k(x) + V(x)\phi_k(x) = \lambda_k \phi_k(x) \tag{4}$$

and $\bar{\phi}$ denotes the complex conjugate of ϕ. Equation (4) has the same structure as the time-independent Schrödinger equation. However, because of the exponential term in Eq. (3), modes with $\lambda_k N >> 1$ essentially do not contribute to Green's function. This is the so-called ground state dominance (GSD). Of interest are the solutions with the lowest values of λ.

The Kronig-Penney model is directly solvable by means of the transfer matrix method [6]. In each cell α between two δ-functions, the free solution of Eq. (4) has two components, with prefactors to be determined. We combine these two prefactors into a two-dimensional vector $\mathbf{a}_\alpha$. The solutions between neighboring cells have to obey the standard conditions of being continuous and to have a kink in between, imposed by the delta function. As a result, the coefficients are related by a set of linear equations $\mathbf{a}_{\alpha+1} = T\mathbf{a}_\alpha$, where T is the transfer matrix. We recall here the form of T valid for repulsive ($\epsilon > 0$) interactions [6]:

$$T = \begin{pmatrix} \cos(\kappa\xi) & \sin(\kappa\xi) \\ -\sin(\kappa\xi) + 2\gamma\cos(\kappa\xi) & \cos(\kappa\xi) + 2\gamma\sin(\kappa\xi) \end{pmatrix} \tag{5}$$

Here we set $\kappa = \sqrt{\lambda}$ for $\epsilon > 0$ and introduce the parameters $u = \kappa\xi$, $E = \epsilon\xi$ and $\gamma = E/2u$. This removes all the ξ dependence from the calculations. In the attractive case ($\epsilon < 0$), only $\lambda < 0$ has to be considered, and we set $\kappa = \sqrt{-\lambda}$ [10]. The eigenvalues $t_\pm$ of the transfer matrix T are given by

$$t_\pm = f \pm \sqrt{f^2 - 1} \tag{6}$$

where f stands for $f(E, u) = \cosh(u) + (E/2u)\sinh(u)$ when $\epsilon < 0$ and for $f(E, u) = \cos(u) + (E/2u)\sin(u)$ when $\epsilon > 0$. In an infinite, periodic array, real $t_\pm$ values imply in general eigenfunctions of unbounded growth. For physical reasons, one has from Eq. (6) $f^2 \leq 1$, which determines the allowed values of λ (band condition). Each band is uniquely parametrized by continuous k values in the interval between $-\pi$ and π. For a given value of E, the dispersion relation between λ and k is $\cos k = f(E, u(\lambda))$, the form which replaces the free solution $\lambda = k^2$ for $E = 0$. Because of GSD, we seek a solution in the neighborhood of the lowest band edge λ_g, and introduce $\Delta\lambda = \lambda - \lambda_g$. Expansion of the dispersion relation to leading order yields

$$\Delta\lambda \simeq \frac{1}{2}(l_{\text{eff}}/\xi)^2 k^2 \tag{7}$$

Here the effective statistical segment length l^2_{eff} is given by

$$l^2_{\text{eff}} \simeq \frac{4u_g^3}{(-E)(u_g + \sinh u_g)} \quad \text{and} \quad l^2_{\text{eff}} \simeq \frac{4u_g^3}{E(u_g + \sin u_g)} \tag{8}$$

for the attractive and the repulsive case, respectively. Note the relation $u_g = \sqrt{|\lambda_g|}$. In the limiting situations $|E| >> 1$ and $|E| << 1$, a simple analytic solution for λ_g can be given [10]. Generally, we have $l^2_{\text{eff}} \leq l^2$, i.e., the chain is always squeezed in the periodic potential. Using Eq. (3) and the expression for $\Delta\lambda(k)$, normalized Green's function can be written as

$$G(x, x'; N) \simeq (2\pi l^2_{\text{eff}} N)^{-1/2} \exp\left\{-\frac{1}{2}\frac{(x - x')^2}{l^2_{\text{eff}} N}\right\} \tag{9}$$

Hence, a long polymer chain behaves in a periodic potential as a free chain endowed with an effective statistical segment length. However, this ground state approximation holds only for chains whose lengths exceed the cross-over value $N_c \sim \xi^2/l^2_{\text{eff}}$.

Localization at isolated defects

We consider now the situation of a single defect site, where two barriers are at the distance ξ' instead of ξ and the deviation from periodicity is thus $\delta = (\xi' - \xi)/\xi$. In this case, there may appear solutions for λ_L outside the bands (so-called defect levels). Again, because of GSD, only defect levels below the lowest band are of interest. The defect states correspond to real eigenvalues $t_\pm$ of Eq. (6) and are localized; their localization length L is given by $L = 1/\ln t_+$. Now, real solutions exist only if the defect (represented by a matrix T') "switches" between a growing (t_+) and a decaying (t_-) solution. This condition can be written in the form [10]

$$\mathbf{l}_+(u_L, E)T'(u_L, E, \delta)\mathbf{r}_+(u_L, E) = h(u_L, E, \delta) \equiv 0 \tag{10}$$

where $\mathbf{l}_+$ and $\mathbf{r}_+$ are the left and the right eigenvectors of T. In the repulsive case, positive δ lead to defect levels below λ_g, whereas the opposite is true in the attracting case; hence

$E\delta > 0$ must be obeyed for defect levels below λ_L. In general, $\lambda_L(E,\delta)$ can be obtained numerically. However, for $0 \leq E\delta << 1$, an analytical solution exists [10]. Denoting now the difference between the lower band edge and the defect level by $\Delta\lambda = \lambda_g - \lambda_L$, we have

$$\Delta\lambda \simeq l_{\text{eff}}^2 \frac{1}{2}\left(\frac{E\delta}{2}\right)^2 \tag{11}$$

and as localization length

$$L/\xi \simeq \frac{2}{E\delta} = \frac{2}{\epsilon(\xi' - \xi)} \tag{12}$$

Now the chain will be localized at the defect if its length is much larger than the cross-over value N_L, where N_L again follows from GSD, $N_L\Delta\lambda \sim 1$. Hence defects of the periodic structure can act as adsorbing sites for the chain and the usual adsorption formalism for polymers may be applied [11]. It is further interesting to note the asymmetry between the case $\xi > \xi'$ and the case $\xi < \xi'$. For instance, in the repulsive case, the removal of a barrier from the regular array results in an *adsorbing* defect; the latter can localize the chain and hence seriously influence its statistical properties. On the other hand, insertion of a barrier creates a repulsive defect which disturbs the chain conformations only slightly.

Finite defect concentrations - unusual behavior of polymers in disordered media

We now turn to the case of a finite concentration ρ of defect levels. In particular, we study the dependence on ρ of the average squared end-to-end distance R^2 of the chain [12]. We assume adsorbing defects, i.e., $E\delta > 0$. Furthermore, we take the chain length N to be much larger than the cross-over value N_L. At very low ρ, the chain will seldom get adsorbed by defects and will spend most of its time in the defect-free region. Once trapped, its escape probability, p_e, is

$$p_e \simeq e^{-\Delta F/kT} \simeq e^{-\Delta\lambda N} \tag{13}$$

where ΔF denotes the difference in Helmholtz (free) energies between the adsorbed and the free conformations. Observed over a sufficiently long time period, R^2 is thus given by $R^2 = g(\rho)R_0^2 + (1 - g(\rho))L^2$, where $g(\rho)$ denotes the probability for the chain to be free at a given ρ and $R_0^2 = l_{\text{eff}}^2 N$ for the defect-free region. Clearly, $g(\rho)$ is monotonic, with $g(0) = 1$, $g(1) = 0$ and, because of detailed balance, $g(\rho)/(1-g(\rho)) = p_e$. Note that defects overlap strongly when their distance is less than L, so that L may be regarded as the eigenvolume (width) of a defect. Thus for $\rho L << p_e$, the ratio $R^2(\rho)/R_0^2$ is a rapidly decreasing function of ρ with a negative slope proportional to $1/p_e$, i.e., very large for $\rho \to 0$ (Fig. 1). When ρL is larger than p_e, the chain is mostly localized at the defects ($R^2 \simeq L^2$).

So far we have assumed that the chain cannot bridge two or more defects. However, even for $\rho << 1/L$, a sufficiently long chain may form bridges between defects, which then allow the chain to extend. We denote by ρ_c the cross-over value where bridging starts and hence $R^2(\rho)$ begins to increase again. In order to estimate ρ_c, we consider two defects at a distance X. Each defect can be again modeled as an adsorbing δ-function in an effective environment. The adsorptive strength Γ of a defect is taken to be $\Gamma = 1/L$;

this leads to one localized state separated from the desorbed states by $\Delta\lambda$ (see Eq. (11)). The corresponding equation for the defect states ϕ_L has then the form

$$\frac{1}{2}l_{\text{eff}}^2\Delta\phi_L(x) + \lambda\phi_L(x) + \Gamma l_{\text{eff}}^2 \sum_k \delta(x - x_k)\phi_L(x) = 0 \tag{14}$$

where one sums over the defects. The two defects considered are a two-level system whose lower level corresponds to bridging. Since we assume $N >> N_L$, we expect a solution for $\rho_c << \Gamma = 1/L$. The difference between the two levels due to coupling is $\Delta\lambda_2 \simeq (1/2)\Gamma^2 l_{\text{eff}}^2 \exp(-\Gamma X)$. Consider now the chain fixed at one defect (quenched case). Again, because of GSD, the bridging contribution from the next defect dominates if $\Delta\lambda_2 N > 1$. Hence, for $\rho > \rho_c \simeq 1/\ln(l_{\text{eff}}^2 N/2L^2) \sim 1/\ln(N)$, the chain is more extended than at lower defect densities. However, if the chain is not fixed, it can minimize its Helmholtz energy by moving in the disordered environment (viewed from the chain, this is an annealed disorder) [13]. The chain will prefer defects which are close to each other. On the other hand, the chain will need rather long times to find such places. To estimate the cross-over value ρ_c for bridging, we calculate the Helmholtz energy gain per chain $< N\Delta\lambda_2 >$, averaged over the distribution of distances between two defects $P(X) = \rho\exp(-\rho X)$. Events for $X << L$ do not lead to an extension of the chain. However, restricting the average over distances X larger than $X_0 \simeq L$ only yields a constant factor of the order of unity. When the averaged Helmholtz energy gain $< N\Delta\lambda_2 >$ is of the order of kT (i.e., unity), bridging should dominate the chain conformations. Using the relation $N_L = 2L^2/l_{\text{eff}}^2$, we get for $N >> N_L$

$$\rho_c L \simeq \frac{N_L}{N} \sim 1/N \tag{15}$$

Note that near $\rho = 1$ (so that $1 - \rho << 1$), the situation differs considerably from that near $\rho = 0$ since the ξ cells are now repulsive defects. As discussed above, in the case of very diluted repulsive defects, the chain conformations are only weakly perturbed. This results in a clear-cut asymmetry of the function $R^2(\rho)$, as also obtained in Fig. 1. That $R^2(\rho)$ displays a minimum can be explained by a simple argument. Since only two lengths ξ and ξ' come into play, both limiting situations, $\rho = 0$ and $\rho = 1$, are periodic and therefore lead to extended conformations (cf. Bloch's theorem). On the other hand, the region in between is dominated by disorder (see also [13, 14, 15]), and hence shows localized states which lower $R^2(\rho)$. $R^2(\rho)$ is an increasing function of ρ for $\rho_c < \rho < 1$. For finite chains, there exists additionally the region $0 < \rho < \rho_c$ where single adsorptive defects determine the chain extension. Note that from Eq. (15) it follows that for $N \to \infty$ one has $\rho_c \to 0$ and thus the minimum shifts to zero. There appears a jump (singularity) at $\rho = 0$ for $N \to \infty$. Hence, for infinite chains, only the increasing behavior (disorder effect) for $\rho > 0$ remains. The minimum is a cross-over effect. It is worth noting that the effects discussed here are independent of the underlying periodic background. The role of the latter is included in l_{eff}, Eq. (8), and can in fact be reflected by taking simple voids between the defects, which yields $l_{\text{eff}} = l$ in Eq. (14).

Monte Carlo simulations

The results of the last section were obtained by focusing on the cross-over behavior of a chain bridging between two attracting defect sites. Qualitatively, we have established

two basic features, namely the strong asymmetry between the situations for $\rho << 1$ and for $1-\rho << 1$ and the fact that the extension of the chain may have its minimum at relatively low ρ_c, i.e. at $\rho_c < 0.5$. To give additional support to our findings, we present now Monte Carlo simulations of single ideal chains in 1D, subjected to aperiodic attractive disorder. We use a striped variant of the well-known bond fluctuation model (BFM) [16] in one spatial dimension, without any excluded volume constraints. The chain consists of N repeat units placed on a 1D grid. The bonds between two repeat units can take all integer values between $-b$ and b. Note that zero values are allowed. It turns out that already $b = 1$ is sufficient to lead to Gaussian statistics for the unperturbed chains reasonably fast. The random jump dynamics for the chain repeat units is now generated as follows. In each attempted monomer move, a repeat unit and a jump direction (± 1 on the 1D lattice) are chosen at random. The repeat unit must again fulfil the bond constraints at its new position. In an external potential, the motion is additionally biased via the standard Metropolis prescription which ensures detailed balance. A Monte Carlo step (MCS) is defined as N attempted monomer unit moves, i.e., during one MCS each monomer has on average one chance to move. This coarsely grained polymer model reproduces the equilibrium properties correctly and leads to realistic diffusional dynamics for times larger than a few MCS [17]. The disorder is taken into account by distributing the traps randomly, in an uncorrelated manner, over the lattice. To reduce the simulation time, we always let the initial chain extend over at least one trap (one trap is always located at the position of the first monomer unit when the chain is created). Each chain is then allowed to relax during 100 000 MCS to remove all statistical artefacts from the chain generation process. Note that the chain can escape from the initial adsorbing site but that in 1D it frequently returns back. On the other hand, if the chain is placed far away from such a site (consider the very dilute limit $\rho \to 0$), it needs a very long time to find it. For finite simulation times, our prescription underestimates $\langle R_g^2 \rangle$ for $\rho \to 0$ (since we always have a trap).
In Fig. 1, we plot the squared radius of gyration $\langle R_g^2 \rangle$ vs. ρ for $N = 80$. The point $\rho = 0^+$ corresponds to a single adsorbing site. The adsorption strength of each defect is chosen as $\epsilon = 0.223$, which corresponds to an escape probability of 0.8 for an individual repeat unit. Note that, as predicted, the curve is not symmetric in the limiting situations, $\rho << 1$ and $1-\rho << 1$. The minimum of $\langle R_g^2 \rangle$ lies near $\rho \simeq 0.3$. Note further the very steep descent of $\langle R_g \rangle(\rho)$ for small ρ (remember that $\langle R_g^2 \rangle(0)$ is underestimated), in contrast to the behavior for ρ near unity.

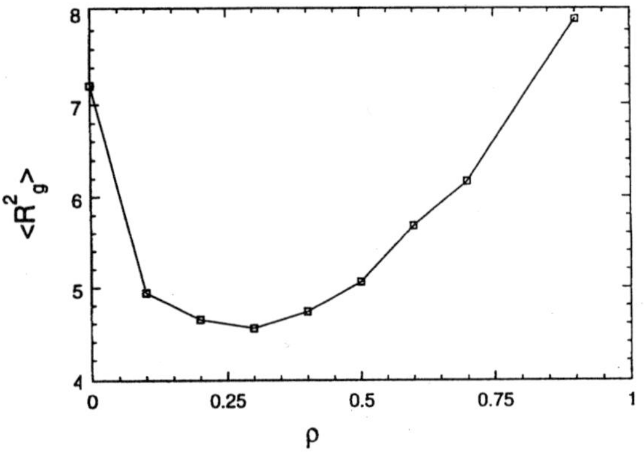

Fig. 1: Radius of gyration vs. the defect density ρ. The data of 100 independent chains comprising 80 repeat units ($N = 80$) are accumulated for $2 \cdot 10^6$ MCS. A pronounced minimum is obtained for $\rho \simeq 0.3$.

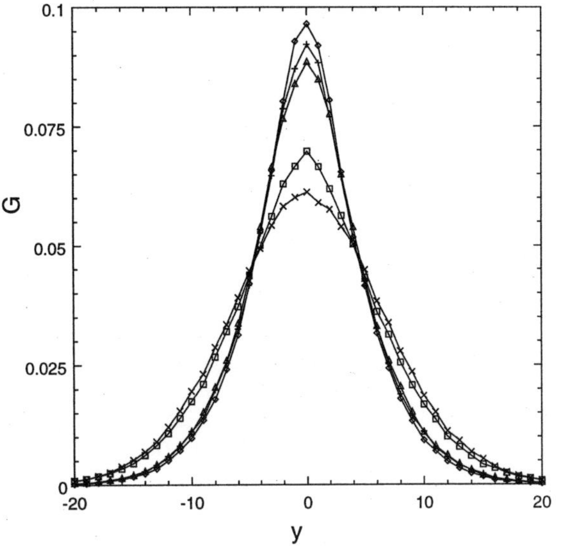

Fig. 2: Distribution function of the end-to-end distance $G(x, x', N = 80)$ vs. the relative distance $y = x' - x$ in lattice units. $\rho = 0^+(\square)$, $0.1(+)$, $0.3(\diamond)$, $0.5(\triangle)$ and $1.0(\times)$.

To stress the non-monotonic behavior of the chain extension with ρ, we display in Fig. 2 normalized Green's function $G(x, x', N = 80)$ (probability distribution of the chain end-to-end distance) for five different values of ρ. The non-monotonic behavior of $G(x', x)$ can be nicely seen.

Diffusion of chains under disorder

So far we have investigated the behavior of ideal chains embedded in layered host materials, as frequently realized by structured polymer systems such as diblock copolymers or liquid crystalline polymers. Simplifying the external potential through a Kronig-Penney model, we succeeded in displaying an exact solution for the perfectly periodic case. Now, isolated defects can adsorb the chain, yielding localized states, whose Helmholtz energy lies below that of the extended (Bloch) chain conformations. In the presence of several adsorbing defects, the chain can bridge between them, and thus extend again. As a consequence, the radius of gyration of the chain depends in a non-monotonic fashion on ρ. In all this, the role of the periodic background is marginal. The main results of our approach are that there is an asymmetry in $\langle R_{\mathrm{g}}^2 \rangle$ as a function of ρ (due to the different roles of repulsive and attractive defects) and that $\langle R_{\mathrm{g}}^2 \rangle$ shows a pronounced minimum for a value of ρ smaller than 1/2. Both effects are supported nicely by Monte Carlo simulations.

An interesting open problem concerns the diffusion of polymer chains in such layered structures. However, the transformation of the problem of the polymer *dynamics* in 1D to a quantum mechanical model is not obvious. Moreover, the external potential (periodic or aperiodic) is always a non-linear function of the monomeric unit position. Hence the full Fokker-Planck formalism must be applied, and we recall that determining the dynamics of polymer chains even in perfect periodic potentials is a very hard task. On the other hand, there exists another physical model which leads to a dynamic

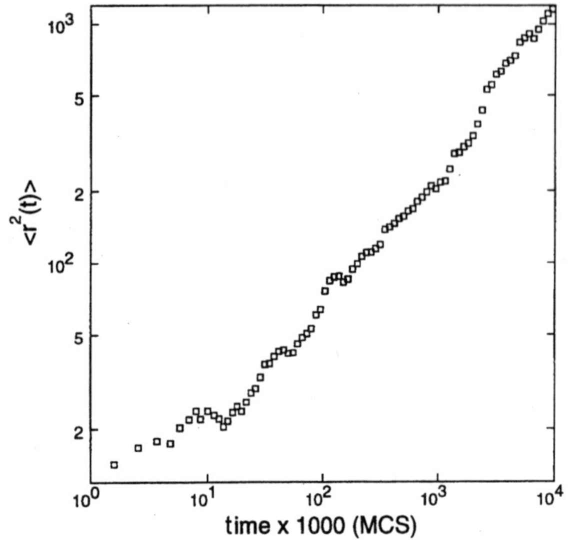

Fig. 3: The mean square displacement averaged over all monomer units and over 100 chains as a function of the simulation time ($N = 80$, $\rho = 0.32$).

problem akin to that of a polymer moving in a 1D disordered medium, namely the analysis of the motion of interfacial lines in a 2D solid-on-solid growth model, as recently discussed by Nieuwenhuizen [8]. Within this approach, the moving interface line is equivalent to a directed random walk in 2D which, in turn, is no more than the graph of a 1D random walk over the monomeric unit positions along the chain.

However, from Monte Carlo simulations, we can obtain valuable information about the dynamic behavior of chains in disordered environments. A first observation is that the chains hardly ever reach diffusive behavior within reasonable simulation times, even if the disorder is not very strong. In Fig. 3, we show $< r^2(t) >$, the mean squared displacement of the monomer units, $< r^2(t) >= (1/N) \sum_{k=1}^{N} < r_k^2(t) >$, as a function of time. The results are averaged over all monomer units within the chain.

Note that within the time scale of 10^7 MCS, a chain of length $N = 80$ is still subdiffusive. In the broad range between 10^4 and $5 \cdot 10^6$ MCS, the slope in Fig. 3 is approximately 1/2. Here the chain explores different areas of the disorder, still finding regions with higher densities of traps and hence with lower local Helmholtz energies [8]. As a consequence, escaping such regions becomes increasingly difficult, thus resulting in subdiffusive behavior. Note that the corresponding time scale is of the order of 10 to 100 Rouse times for the unperturbed chain; the slope of $\sim 1/2$ should not be related to Rousian diffusion. From that it follows that we cannot define a diffusion coefficient for all values of ρ (diffusive behavior is found, however, for very small and very large ρ values). As a measure of the mobility of the chains, we hence use the value $< r^2(\tau) >$ for $\tau = 2 \cdot 10^6$ MCS. This is the longest time used for all systems considered. The results are displayed as a function of ρ in Fig. 4.

One should note that this *dynamic* quantity again shows a pronounced non-monotonic behavior as a function of ρ, the fact qualitatively similar to the static results. However, here the differences between small ($\rho \simeq 0$), large ($\rho \simeq 1$) and moderate ρ values are much larger than those displayed by the static results (note the logarithmic scale in Fig. 4). Our dynamic results are quite similar to those recently reported by Slater and

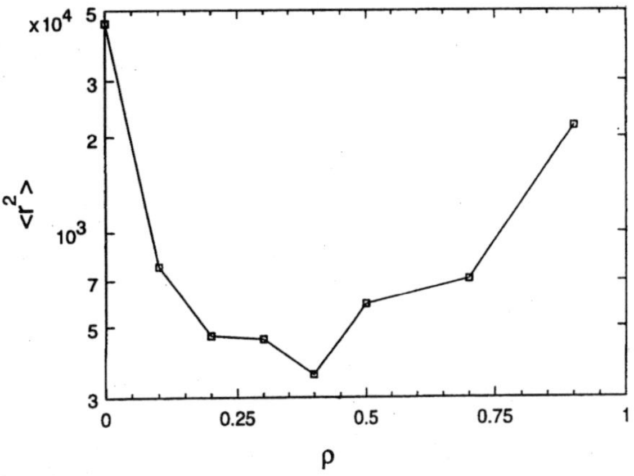

Fig. 4: The mean square displacement of monomer units at $t = 2 \cdot 10^6$ MCS as a function of the defect density ρ ($N = 80$).

Wu [18] from 2D Monte Carlo simulations; this may lead to interesting consequences for the electrophoretic mobility of chains in layered structures.

Acknowledgments

Support by the DFG and by the Fonds der Chemischen Industrie is gratefully acknowledged.

References

[1] P. G. de Gennes, Solid State Phys. Suppl. **14**, 1 (1978).
[2] W. Renz and M. Warner, Phys. Rev. Lett. **56**, 1268 (1986).
[3] J.-U. Sommer and M. Daoud, Europhys. Lett. **32**, 407 (1995).
[4] J.-U. Sommer and M. Daoud, Phys. Rev. E 53, 905 (1996).
[5] J.-U. Sommer, A. Halperin, and M. Daoud, Macromolecules **27**, 6991 (1994).
[6] J. M. Ziman, *Models of Disorder*, Cambridge University Press, Cambridge, 1979.
[7] R. de L. Kronig and W. Penney, Proc. R. Soc. (London), Ser. A **130**, 499 (1931).
[8] T. M. Nieuwenhuizen, Phys. Rev. Lett. **78**, 3491 (1997).
[9] M. Doi and S. Edwards, *The Theory of Polymer Dynamics*, Clarendon Press, Oxford, 1986.
[10] J.-U. Sommer and A. Blumen, J. Chem. Phys. **105**, 6008 (1996).
[11] P. G. de Gennes, Macromolecules **14**, 1637 (1981).
[12] J.-U. Sommer and A. Blumen, Phys. Rev. Lett **79**, 439 (1997).
[13] M. E. Cates and R. C. Ball, J. Phys. (Paris) **49**, 2009 (1988).
[14] S. F. Edwards and M. Muthukumar, J. Chem. Phys. **89**, 2435 (1988).
[15] J. Machta and R. A. Guyer, J. Phys. A **22**, 2539 (1989).
[16] I. Carmesin and K. Kremer, Macromolecules **21**, 2819 (1988).
[17] W. Paul and K. Binder, Europhys. Lett. **6**, 585 (1988).
[18] G. W. Slater and S. Y. Wu, Phys. Rev. Lett. **75**, 164 (1995).

Wetting Reversal Transition in Phase-separated Polymer Mixtures

Jan Genzer,*† Jakob Heier and Edward J. Kramer

Materials Department, University of California at Santa Barbara, Santa Barbara, California, 93106-5050, USA

Abstract: We investigate a wetting reversal transition in thin films of two-phase mixtures of poly(ethylene-propylene) (PEP) and its deuterated analogue (dPEP) on a substrate covered by a self-assembled monolayer (SAM) whose surface energy, γ_{SAM}, is tuned by varying the SAM composition. As γ_{SAM} increases from 21 to 24 mJ/m^2, a transition from a three-layer (air/dPEP/PEP/dPEP/SAM) to a two-layer (air/dPEP/PEP/SAM) structure occurs at increasing T_c - T, where T_c and T are the critical and transition temperatures, respectively. As the system structure changes from three-layer to two-layer, the thicknesses of the dPEP-rich wetting layers at the air/mixture and mixture/SAM interfaces are found to smoothly increase and decrease, respectively, while the thickness of the PEP-rich layer (ca. one half of the total film thickness) does not change. The dependence of the transition temperature on γ_{SAM} is predicted by a simple model using the experimental data on the surface energies of PEP/dPEP and estimates of the interfacial energy between PEP and dPEP.

INTRODUCTION

The behavior of phase-separated polymer blends has been subject of much theoretical and experimental work in the past few years. Upon quenching into the unstable region of the phase diagram, critical mixtures of A and B polymers phase-separate into A-rich and B-rich coexisting phases. In the bulk, the concentration fluctuations that govern the phase separation process are random. As a result, the final morphology consists of interconnected A-rich and B-rich domain structures which coarsen slowly with time (Ref. 1). However, in thin films, the situation changes drastically. The presence of additional interfaces, i.e. the air/mixture and mixture/substrate, causes the directions of the compositional waves in the mixture close to the interfaces to be modified such that the resultant domains are oriented parallel to these interfaces (Refs. 2-6). Hence the phase morphology and its time evolution in thin, phase-separated polymer films are governed by an interplay between phase separation processes and interactions of the A-rich and B-rich polymer phases with the air and with the substrate. While the air surface will always prefer a low surface energy (say, the A-rich phase), the substrate interfacial preference can change from the A-rich phase to the B-rich phase if the

† Present address: Department of Chemical Engineering, North Carolina State University, Raleigh. North Carolina, 27695-7905, USA.

 CCC 1022-1360/99/$ 17.50+.50/0

nature of the substrate surface changes (Refs. 4-6) or, for a constant substrate surface, as the composition of the coexisting phases changes (Ref. 2). Thus if the properties of the substrate surface can be tuned sensitively, it may be possible to observe a transition from an A-rich/B-rich/substrate (two-layer) structure to an A-rich/B-rich/A-rich/substrate (three-layer) structure as the polymer mixture is cooled from the critical temperature. Recently, we experimentally demonstrated that this possibility can be realized using phase-separated mixtures of poly(ethylene-propylene) (PEP) and its deuterated analogue (dPEP) deposited on substrates covered with self-assembled monolayers (SAM) composed of blends of alkanethiols with hydrophobic and hydrophilic end groups (Ref. 7). In this work we present a summary of our recent results and extend our previous investigations to explore the dependence of the thickness of the dPEP-rich wetting layer at the mixture/SAM interface on γ_{SAM} in the vicinity of the three-layer to two-layer transition.

EXPERIMENTAL

Smooth and well characterized substrates are a necessary prerequisite for a precise control of the interactions at the mixture/substrate interface. Such substrates can be fabricated by depositing SAMs of end-functionalized alkanethiols onto gold (Ref. 8). If two thiols having different end groups are coadsorbed onto the Au surface, γ_{SAM} is related to the ratio of the two thiols in the SAM blend. The extent to which the SAM surface energy changes depends on the character of the thiol end groups. When the difference between the polymer/SAM interactions for both phase-separated phases is not substantial, such as in isotopic polymer mixtures, one expects that by varying the surface energy of SAM, one can induce either the A-rich or the B-rich phase to wet the substrate. The substrates for the SAMs were produced by evaporating a layer of Au (ca. 1500 Å thick) onto Si wafers covered with a layer of Cr (ca. 150 Å thick). The latter served as an adhesion promoter for Au. The SAMs were prepared by exposing the Au-covered Si wafers to tetrahydrofuran solutions of $HS(CH_2)_{15}COOH$ / $HS(CH_2)_{17}CH_3$ blends for 24 h at room temperature. After adsorption, the SAMs were washed in tetrahydrofuran and dried with nitrogen. By varying the mole fraction, x, of $HS(CH_2)_{15}COOH$ in the $HS(CH_2)_{17}CH_3$ / $HS(CH_2)_{15}COOH$ blend from 0 to 1, γ_{SAM} increased from ca. 20 to 81 mJ/m^2 (Ref. 7). The values of γ_{SAM} were determined using the geometric mean approximation (GMA) from the measurements of the advancing contact angles, θ, of deionized water and methylene iodide (Ref. 9). The applicability of GMA was checked for by comparing the values of γ_{SAM} with the critical surface energies of SAMs, $\gamma_{SAM,c}$. The latter values were obtained from Zisman plots, which were generated by measuring θ using a series of homologous alkanes (Ref. 7). Our results showed that for $\gamma_{SAM} < 25$ mJ/m^2, the differences between γ_{SAM} and $\gamma_{SAM,c}$ were smaller than 1 mJ/m^2.

The PEP and dPEP polymers were prepared by hydrogenating and deuterating, respectively, anionically polymerized 1,4-polyisoprenes (Ref. 10). The degrees of polymerization of both polymers, N, were ca. 2280, which leads to an upper critical solution temperature, T_c, of ca. 93 °C (Ref. 11). Thin films (ca. 4000 Å thick) of critical dPEP/PEP mixtures (50/50 v/v) were spin-coated from toluene solutions onto glass microscope slides, floated onto a bath of deionized water and picked up with the SAM-covered substrates. The samples were annealed in vacuum at four different temperatures below T_c, namely at 23, 44, 66, and 75 °C, for various times ranging from 4 days (75 °C) to 3 weeks (23 °C). The annealing times were chosen such that the samples reached equilibrium (Ref. 12). After annealing, the samples were quenched by immersion into a bath of liquid nitrogen to preserve the structure. Conventional forward recoil spectrometry (FRES) was then used to measure the volume fraction profiles of dPEP and PEP in the samples (Ref. 13).

RESULTS AND DISCUSSION

Figures 1 and 2 show the volume fraction profiles of dPEP (closed circles) and PEP (open circles) in critical dPEP/PEP mixtures on SAMs made of pure $HS(CH_2)_{17}CH_3$ and pure $HS(CH_2)_{15}COOH$, respectively, annealed at 44 °C for 110 h. The volume fractions of dPEP in the dPEP-rich and the PEP-rich coexisting phases are 0.710 and 0.315, respectively (Ref. 12). Figures 1 and 2 demonstrate that in both samples, a dPEP-rich phase is present at the air/mixture interface, in agreement with previous experiments (Refs. 4,12,14). This behavior is expected because the surface energy of dPEP is lower than that of PEP (Ref. 14). However, depending on the substrate, either a dPEP-rich or a PEP-rich phase is present at the mixture/SAM interface. Namely, for SAMs with $x \leq 0.28$, the mixture/SAM interface is wet by the dPEP-rich phase and for SAMs with $x \geq 0.30$, the mixture/SAM interface is wet by the PEP-rich phase. The transition from the three-layer structure to the two-layer structure thus occurs for SAMs with $0.28 < x < 0.30$ which corresponds to $22.4 < \gamma_{SAM} < 23.5$ mJ/m^2. The crossover from the three-layer to two-layer structure was monitored at three additional annealing temperatures below T_c. The results are presented in the form of the phase diagram of the three-layer to two-layer transitions shown in Fig. 3.

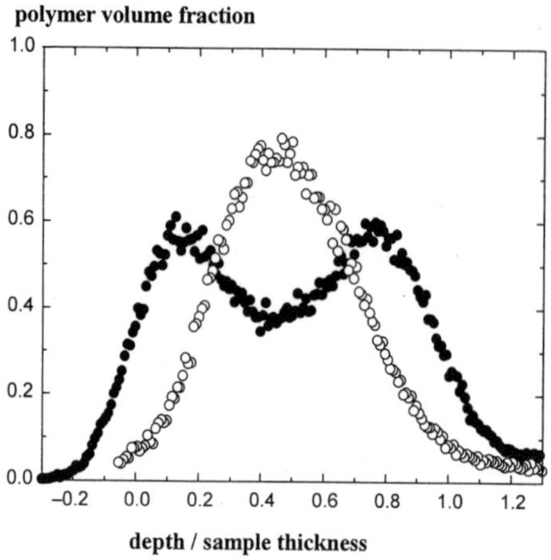

Figure 1 Volume fraction profiles of dPEP (●) and PEP (O) in critical dPEP/PEP mixtures ca. 4000 Å thick at CH_3-terminated SAM annealed at 44 °C for 110 h

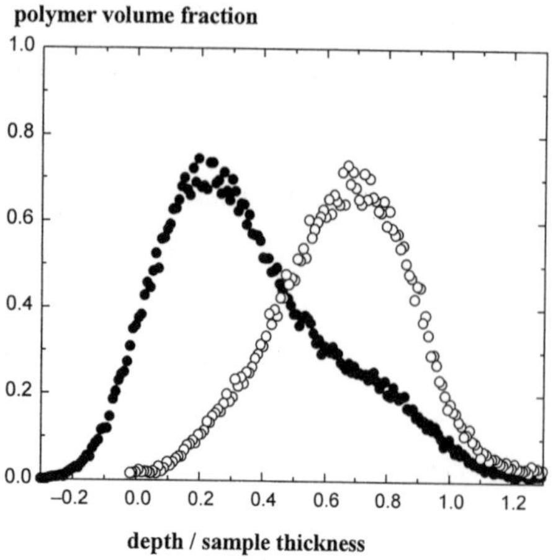

Figure 2 Volume fraction profiles of dPEP (●) and PEP (O) in critical dPEP/PEP mixtures ca. 4000 Å thick at COOH-terminated SAM annealed at 44 °C for 110 h

In Fig. 3, we plot γ_{SAM} as a function of the annealing temperature. The open and solid circles represent the three-layer and two-layer structures, respectively, as determined from the FRES experiments. The solid line is a guide to the eye. The results in Fig. 3 show that with the decreasing quench depth ε (=1-T/T_c), the three-layer to two-layer transition takes place at lower values of γ_{SAM}.

Previously we showed that the transition between the three-layer and two-layer structures takes place when

$$\gamma_{SAM/PEP} - \gamma_{SAM/dPEP} = \gamma_{PEP/dPEP} \qquad (1)$$

where $\gamma_{SAM/dPEP}$ and $\gamma_{SAM/PEP}$ are the interfacial energies of the dPEP- and PEP-rich phases and the SAM, respectively, $\gamma_{PEP/dPEP}$ is the interfacial energy between the dPEP- and PEP-rich phases (Ref. 7). We have also demonstrated that γ_{SAM} at which the three-layer to two-layer transition occurs can be estimated from a simple model based on the combination of Young-Dupré and Good-Girifalco equations (Refs. 15,16):

$$\gamma_{SAM} = \{[(\gamma_{PEP} - \gamma_{dPEP}) - (\gamma_{SAM/PEP} - \gamma_{SAM/dPEP})] / [2 (\sqrt{\gamma_{PEP}} - \sqrt{\gamma_{dPEP}})]\}^2 \qquad (2)$$

where γ_{PEP} and γ_{dPEP} are the surface energies of the PEP- and dPEP-rich phases. From Eq. (1), $\gamma_{SAM/PEP}$ - $\gamma_{SAM/dPEP}$ can be substituted for by $\gamma_{PEP/dPEP}$ and the latter term evaluated using the expression suggested by Tang and Freed (Ref. 17):

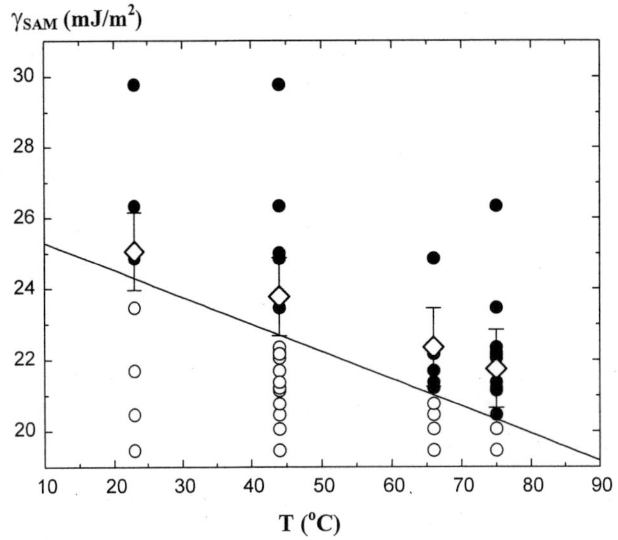

Figure 3 Phase diagram of three-layer to two-layer transitions at the dPEP/PEP mixture/SAM interface for mixtures annealed at four different temperatures below T_c. Also shown are the positions of the three-layer to two-layer transition calculated using the model described in the text and in Ref. 7. (O three-layer structure, ● two-layer structure, ◇ model; the solid line is a guide to the eye)

$$\gamma_{PEP/dPEP} = k_B T a \rho_o (\chi/6)^{0.5} \left\{ 1 - \frac{1.8}{\chi N} - \frac{0.4}{(\chi N)^2} \right\}^{1.5} \tag{3}$$

In Eq. (3), k_B is the Boltzmann constant, T is absolute temperature, *a* is segment length of the polymers (8 Å), ρ_o is the monomer number density (= 0.0077 Å^{-3}), and χ is the interaction parameter on the coexistence curve between dPEP-rich and PEP-rich phases given by $\chi = 0.16/T - 4.6\text{x}10^{-4}$ (Ref. 12).

The open diamonds in Fig. 3 denote the values of γ_{SAM} calculated using Eqs. (2), (3) and $\gamma_{PEP} - \gamma_{dPEP}$ determined from $\Delta\gamma$, the values of the surface energy difference between PEP and dPEP as a function of the composition of PEP/dPEP mixtures reported by Norton and coworkers (Ref. 14). In the calculations we used $\gamma_{PEP} = 30.6$ mJ/m^2 at 25 °C and $-d\gamma/dT \approx 0.052$ mJ/m^2/K (Ref. 18). The error bars denote the uncertainty in the transition values of γ_{SAM} based on the assumption that the values of γ_{PEP} are known to within ±1 mJ/m^2. The agreement between the measurement and the calculation is excellent, particularly at large ε (low T). At smaller ε, the minor deviations between the measured and calculated γ_{SAM} may be caused by the assumption that $\Delta\gamma$ does not change with temperature. Although the variation of $\Delta\gamma$ with temperature is not expected to be large (Ref. 14), it is evident from Eq. (3) that even a small decrease in $\Delta\gamma$ and thus $\gamma_{PEP} - \gamma_{dPEP}$ with increasing temperature could bring the calculated γ_{SAM} closer to the experimental values.

As previously discussed, at a given ε, the system structure depends on the value of γ_{SAM}. It is of interest to investigate the structural changes as the system moves from the three-layer to two-layer configuration. This can be accomplished by depositing thin films of PEP/dPEP mixtures on SAM substrates whose γ_{SAM}'s span a narrow region around the three-layer to two-layer transition. The SAMs were formed by exposing the Au-covered Si wafers to ethanol solutions of $HS(CH_2)_{11}COOH$ / $HS(CH_2)_{11}CH_3$ blends for 24 h at room temperature. By varying the mole fraction, x, of $HS(CH_2)_{11}COOH$ in the $HS(CH_2)_{11}CH_3$ / $HS(CH_2)_{11}COOH$ blend from 0 to 1, γ_{SAM} increased from ca. 19 to 61 mJ/m^2. Thin films of critical dPEP/PEP mixtures were prepared as previously described. The samples were annealed at 41 °C in vacuum for 7 days and conventional FRES was used to measure the polymer volume fraction profiles. Figure 4 shows the variation of the thickness of the dPEP-rich and PEP-rich wetting layers as evaluated using RUMP (Ref. 19) from the experimental FRES data. From the data in Fig. 3, the three-layer to two-layer transition is expected to occur at γ_{SAM} between 23 and 24 mJ/m^2.

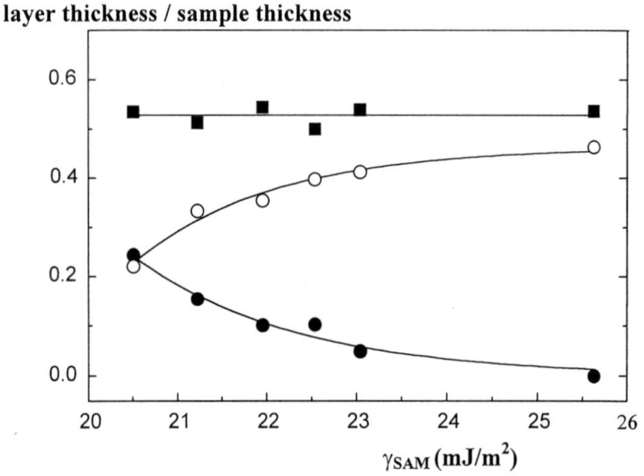

Figure 4 Normalized thickness of the dPEP-rich wetting layers as a function of the surface energy of SAM for samples annealed at 41 °C for 140 h
(● dPEP, mixture/SAM, O dPEP, air/mixture, ■ PEP; the solid lines are guides to the eye)

The results in Fig. 4 show that as γ_{SAM} increases from 20 to 26 mJ/m^2, the thickness of the dPEP-rich layer at the mixture/SAM interface (closed circles) continuously decreases while that of the dPEP-rich layer at the air/mixture (open circles) interface increases. At $\gamma_{SAM} > 23$ mJ/m^2, the system adopts the two-layer structure. Within the resolution of the experiment, the thickness of the PEP-rich wetting layer (closed squares) remains constant over a broad range of γ_{SAM} and is about a half of the total film thickness. This behavior is well understood because the thickness of the PEP-rich layer is fixed by the phase rule. The results shown in Fig. 4 suggest that as γ_{SAM} increases in the vicinity of the three-layer to two-layer transitions, the affinity of dPEP towards the SAM substrate decreases. As γ_{SAM} increases there is a material flow from the mixture/SAM interface across the PEP-rich layer towards the air/mixture interface that builds up the surface dPEP-rich wetting layer.

The main factors that control the thicknesses of the dPEP-rich layers are the strengths of the driving forces governing the segregation of dPEP at the air/mixture and mixture/SAM interfaces. Steiner and Klein recently showed that the growth of wetting layers in thin phase-separated polymer mixtures is driven primarily by long-range surface forces (Ref. 3). The long-range van der Waals energy between two parallel planes is given by $W = -A/(12\pi D^2)$, where A is the Hamaker constant and D is the separation of the planes (Ref. 20). It can be shown (Ref. 21) that within a simple model the thickness of the dPEP-rich layer at the mixture/SAM interface normalized by the total sample thickness, $l_{d,AM}/L$, is given by

$$\frac{l_{d,SAM}}{L} \approx \frac{0.5}{1+\left[\dfrac{\sqrt{\gamma_{PEP}}+\sqrt{\gamma_{dPEP}}}{\sqrt{\gamma_{PEP}}+\sqrt{\gamma_{dPEP}}-2\Phi\sqrt{\gamma_{SAM}}}\right]^{1/3}} \tag{4}$$

where Φ ($\cong$1) is a Girifalco-Good correction factor accounting for the nature of the intermolecular interactions (Ref. 16). Equation (4) demonstrates that (i) as $\gamma_{SAM} \rightarrow 0$, $l_{d,SAM}/L$ approaches 0.25 and (ii) as $\gamma_{SAM} \rightarrow \gamma_{dPEP}$, $l_{d,SAM}/L$ decreases to 0. These trends are in accord with the experimental results. At low γ_{SAM} the affinity of dPEP toward the SAM substrate is comparable to that at the air/mixture interface. As a result, the thicknesses of the dPEP-rich wetting layers at both the air/mixture and mixture/SAM interfaces are approximately equal. Upon increasing γ_{SAM}, the system moves closer to the three-layer to two-layer transition and the attraction of dPEP towards the SAM decreases. In this regime, the thickness of the dPEP-rich layer at the mixture/SAM interface decreases and the excess of material is used to build up the dPEP-rich wetting layer at the air/mixture interface.

CONCLUSION

In summary, we have demonstrated that the ultimate phase morphology in thin polymer films can be controlled by suitably adjusting the polymer/substrate interactions and/or the annealing temperature. This control may be exploited in a number of ways. For example, microcontact printing can be used to fabricate a lateral pattern of compositionally different SAMs on the substrate surface. A phase-separated polymer mixture cast on such surfaces may then produce phase structures that are modulated laterally at one temperature but disappear at another one. One could also explore the phase morphologies that will form if the temperature is changed from above the two- to three-layer transition to below it at various stages in the spinodal decomposition process. It seems likely that interesting new morphologies may result.

ACKNOWLEDGMENT

This research was supported by the Division of Materials Research, NSF Polymers Program, under Grant No. NSF-DMR92-23099. The use of the Central Facilities of the Cornell Materials Science Center for the ion beam analysis funded by the NSF-DMR-MRSEC program is acknowledged. We are greatly indebted to Prof. F. S. Bates (University of Minnesota) for supplying PEP and dPEP, Prof. G. M. Whitesides (Harvard University) for donating $HS(CH_2)_{15}COOH$, Dr. J. Srogl (Emory University) for synthesizing $HS(CH_2)_{11}COOH$, and Dr. B. B. Sauer (DuPont) for providing the values of the surface energy of PEP.

REFERENCES

(1) T. Hashimoto, Chap. 6 in *Materials Science and Technology*, Vol. 12, Ed. E. L. Thomas, VCH, Weinheim, 1993, p. 251.

(2) U. Steiner, J. Klein, E. Eiser, A. Budkowski and L. J. Fetters, *Science* **258**, 1126 (1992); U. Steiner, J. Klein and L. J. Fetters, *Phys. Rev. Lett.* **72**, 1498 (1994).

(3) U. Steiner and J. Klein, *Phys. Rev. Lett.* **77**, 2526 (1996).

(4) F. Bruder and R. Brenn, *Phys. Rev. Lett.* **69**, 624 (1992).

(5) G. Krausch, C.-A. Dai, E. J. Kramer, J. F. Marko and F. S. Bates, *Macromolecules* **26**, 5566 (1993).

(6) G. Krausch, E. J. Kramer, M. H. Rafailovich and J. Sokolov, *Appl. Phys. Lett.* **64**, 2665 (1994).

(7) J. Genzer and E. J. Kramer, *Phys. Rev. Lett.* **78**, 4946 (1997).

(8) For a recent review of applications of self-assembled monolayers, see D. L. Allara, *Biosens. Bioelectron.* **10**, 771 (1995).

(9) B. B. Sauer and N. V. Dipaolo, *J. Colloid Interface Sci.* **144**, 527 (1991).

(10) A. Zirkel, D. Richter, W. Pyckhout-Hintzen and L. J. Fetters, *Macromolecules* **25**, 954 (1992).

(11) M. D. Gehlsen, J. H. Rosedale, F. S. Bates, G. D. Wignall, L. Hansen and K. Almdal, *Phys. Rev. Lett.* **68**, 2452 (1992).

(12) J. Heier, unpublished results.

(13) For description of conventional FRES, see, e.g.: P. Mills *et al.*, *Appl. Phys. Lett.* **45**, 958 (1984).

(14) L. J. Norton, E. J. Kramer, F. S. Bates, M. D. Gehlsen, R. A. L. Jones, A. Karim, G. P. Felcher and R. Kleb, *Macromolecules* **28**, 8621 (1995).

(15) S. Wu, *Polymer Interface and Adhesion*, Marcel Dekker, New York, 1982.

(16) L. A. Girifalco and R. J. Good, *J. Phys. Chem.* **61**, 904 (1957); *ibid.* **64**, 561 (1960).

(17) H. Tang and K. F. Freed, *J. Chem. Phys.* **94**, 6307 (1991).

(18) B. B. Sauer, to be published.

(19) RUMP is a computer program for simulating the FRES spectra. For details, see L. R. Doolittle, *Nucl. Instrum. Methods Phys. Res., Sect. B* **9**, 344 (1985); *ibid.*, **15,** 227 (1986).

(20) J. Israelachvili, *Intermolecular and Surface Forces*, Academic Press, New York, 1992.

(21) J. Genzer and E. J. Kramer, *Europhys. Lett.* **44**, 180 (1998).

Interfacial Tension Between Polymer-containing Liquids – Predictability and Influences of Additives

B.A. Wolf

Institut für Physikalische Chemie der Universität Mainz
Welder-Weg 13, D-55099 Mainz, Germany

Abstract: The first part of the contribution deals with the interfacial tension, σ, of phase-separated polymer solutions in single or mixed solvents and of binary polymer blends as a function of the relative distance to the critical temperature of the system, special attention being paid to the possibilities of theoretical prediction. Two methods are discussed in more detail. One is based on a realistic description of the Gibbs energy of mixing as a function of composition, the second correlates σ with the length of the measured tie line. The second part is devoted to another aspect, namely the effects of additives on the interfacial tension between the coexisting phases of demixed polymer solutions and between highly incompatible polymers. In the former case, it is demonstrated that an addition of a thermodynamically good solvent is normally associated with a reduction in σ; however, adding a high-molecular-weight compound which is incompatible with the dissolved polymer leads to an increase in σ. The interfacial tension between incompatible homopolymers is efficiently reduced by block copolymers consisting of monomeric units which are either identical with or different from those of the homopolymers; in contrast to theoretical expectation, the molecular architecture of the additives seems to be of minor importance only. Random copolymers which are insoluble in the homopolymers can also efficiently reduce the interfacial tension.

INTRODUCTION

Since approximately five years ago, we have been engaged in the study of interfacial phenomena. The present contribution gives a compilation and analysis of the results obtained in the two areas named in the subtitle. The work of the colleagues who have laid the scientific ground for these investigations is cited in the references, which also name the coworkers who carried out the work reported here. All measurements were performed by means of the spinning, sessile and pendant drop methods.

PREDICTABILITY

In view of the relevance of the interfacial tension σ for technical applications of polymers and the fact that accurate measurements of this quantity require considerable experimental scrutiny, it would be very useful to predict σ either from suitable theories or at least from semiempirical relations. We have therefore tried to find ways for such predictions.

 CCC 1022-1360/99/$ 17.50+.50/0

In one approach we have introduced ε, the reduced hump energy, as the area between $\Delta\overline{\overline{G}}/RT$, the reduced segment molar Gibbs energy of mixing as a function of composition and the double tangent to that curve, which fixes the composition of the coexisting phases. According to phenomenological considerations, this quantity should be closely correlated with the interfacial tension[1,2]. The evaluation of the entire accessible experimental material concerning σ and ε of phase-separated polymer solutions indeed results in a scaling law of the form $\sigma = E\varepsilon^F$, where F is close to 0.5 for all systems whereas E varies markedly. Similarly, ε depends on τ, the relative distance from the critical temperature, according to $\varepsilon = A\tau^B$, where B is approximately 2.4 and relatively constant, in contrast to A. According to a more detailed analysis of the situation, the constants A and E of the relations formulated above are not universal[2], contrary to some theoretical expectation.

In spite of this deficiency, the scaling laws provide the means for a rough prediction of σ as a function of temperature from the sole knowledge of the critical point of the system. Using a two-parameter equation to describe the concentration and temperature dependence of the Flory-Huggins interaction parameter, it is possible to calculate the reduced hump energy by means of the critical data. The temperature for which one wants to know σ is then expressed in terms of τ and yields ε, which in turn is converted into σ; notwithstanding the uncertainties in A and E, this prediction is normally better than that based on a direct correlation of (concentration-independent) Flory-Huggins interaction parameters and interfacial tensions.

In an older approach[3,4], we have established a master curve linking σ to Δφ, the length of the corresponding tie line (expressed in terms of volume fractions φ). For phase-separated solutions of polystyrenes and similar macromolecules, the relation reads σ (mN/m) = $0.015\, M^{0.5}\Delta\varphi^{3.85}$, where M is the molar mass of the polymer. The application of the just formulated master curve to molecularly non-uniform polymer samples provides a simple means to grasp the effects of polydispersity[5]. This accounts for the fact that the σ remains finite within a limited interval of negative τ values (i.e., at T > UCST or T < LCST) as a result of the molecular non-uniformity of the polymer. The outlined correlation of σ and Δφ also works with ternary systems, as demonstrated for solutions of poly(acrylic acid) in mixed solvent consisting of water and propan-2-ol. In this case, a much larger range of σ and lengths of tie lines is experimentally accessible and the resulting exponent of Δφ is 4.23.

With polymer blends, it is normally impossible to measure σ close to the critical region due to problems in the attainment of equilibria resulting from high viscosity of such mixtures and very similar densities of the coexisting phases. In order to circumvent these difficulties, we have studied blends of different poly(siloxane)s; these substances exhibit very low glass-transition temperatures and are rather fluid. With them, interfacial tensions could be measured for the first time in the vicinity of the critical temperature[6]. It is possible to describe the observed dependence of σ on τ or $\Delta\varphi$ (as shown in Fig. 1), within experimental error, by both the mean field theory and the Ising-3D model.

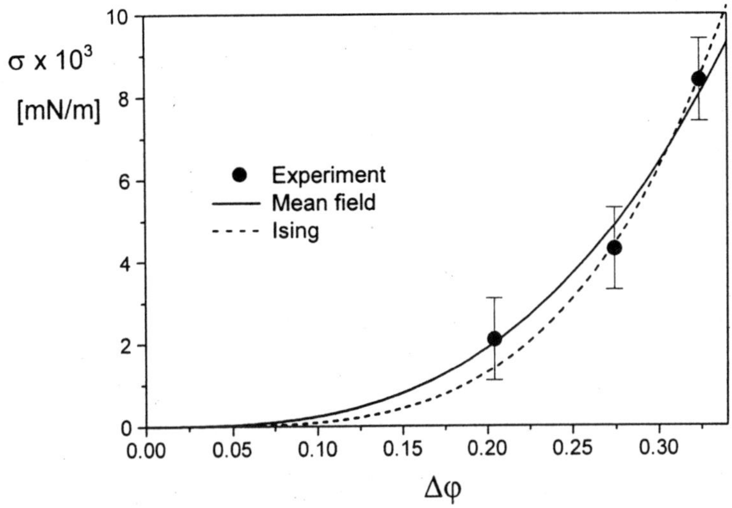

Figure 1: Interfacial tension for mixtures of poly(dimethylsiloxane) (M_w=3 kg/mol) and poly[hexyl(methyl)siloxane] (M_w=113 kg/mol), as obtained by the spinning drop method and theory[7].

ADDITIVES

The sensitivity of the interfacial tension between polymer-containing liquids to small amounts of additives is well established and of technical importance. For a better theoretical understanding of these phenomena, we have studied the effects of low- and high-molecular-weight additives on σ for such binary mixtures of limited mutual solubility.

The effects resulting from an addition of a third component to a phase-separated binary system are normally rather different, depending on its low- or high-molecular-weight nature. An addition of a thermodynamically good (low-molecular-weight) solvent to a demixed polymer solution *reduces* the interfacial tension. Analogously, σ of an incompatible polymer blend becomes smaller as one adds a low-molecular-weight solvent which is completely miscible with both high-molecular-weight components[8]. The situation becomes more complicated if the solvent is good only for one of the polymers but bad for the other; in this case, σ may pass a minimum as the composition of the mixed solvent changes[8].

The addition of a second high-molecular-weight component to a phase-separated polymer solution results in fundamentally different effects. For such systems, it could be shown[9] that the incompatible polymers *increase* the interfacial tension markedly. In terms of the correlation between σ and $\Delta\varphi$ discussed above, this finding can be easily rationalized as an extension of the miscibility gap by the addition of the second polymer, i.e. as a lengthening of the tie line.

As already indicated, measurements with polymer blends are normally possible only far from the region of complete miscibility. We have therefore investigated[10,11] the effects of additives for the system poly(dimethylsiloxane)/poly(ethylene oxide) at temperatures ranging typically from 70 to 150 °C. Two families of block copolymer additives were studied. With type I, the blocks consisted of the monomeric units of the corresponding homopolymers; these additives were varied in their molar mass and molecular architecture. Type II comprised diblock copolymers composed of monomeric units which differ from those of the homopolymers.

In all cases, σ was substantially lowered; for type I additives, the effects range typically from ca. 10 mN/m to 1-4 mN/m when only 0.2 wt.-% is added to the poly(ethylene oxide) phase. The dependence of σ on composition was studied in terms of the x^*_{add}, the base mole fraction of the additive in the droplet phase, i.e. the number of monomeric units contained in the additive divided by the total number of monomeric units. The function $\sigma(x^*_{add})$ can be well described by two parameters: by σ_s, the saturation value of σ for large amounts of the additive, and $x^*_{charact}$, measuring its efficiency. The results demonstrate that the total number of dimethylsiloxane units the type I additives contain is more decisive for their efficiency than their architecture (di-, tri- or multiblocks); this finding is in contrast to theoretical expectation.

One observation deserves special notice. In the cases in which the additive is markedly soluble in both phases, it can make a big difference for the measured interfacial tension to which the substance is added[12]. We are presently investigating this phenomenon in more detail.

The copolymers of type II we have studied are considerably less effective than those of type I as demonstrated in Fig. 2 for an example.

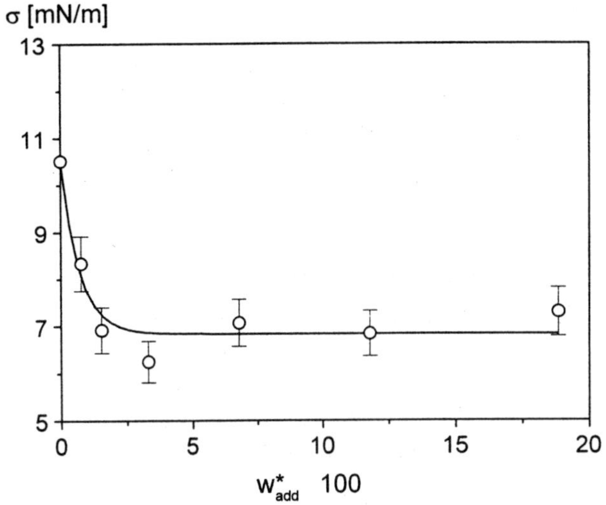

Figure 2: Variation of the interfacial tension at 100 °C between poly-(dimethylsiloxane) (M_w=177 kg/mol) and poly(ethylene oxide) (M_w=41 kg/mol) upon the addition of a diblock copolymer of styrene (75 monomeric units) and methyl methacrylate (110 monomeric units). (w_{add}* is the base mole fraction of the additive in the poly(ethylene oxide) phase).

In view of some drawbacks of block copolymers (like their tendency to form micelles or their slow migration to the interface) and of theoretical interest, we have also begun[13] to study the efficiency of random copolymers. This work was performed with a highly incompatible blend of poly[methyl(phenyl)siloxane] and poly(dimethylsiloxane) where the random copolymer contained 11 mol-% of the former and 89 mol-% of the latter monomeric units. Approximately 1 wt.-% of the additive reduces σ from ca. 2.2. to 1.6 mN/m. It is interesting to note that two different kinetic processes are observed during equilibration of the system. They are tentatively ascribed to the interdiffusion of the components in the vicinity of the phase boundary and to the hydrodynamic relaxation of the droplet. Concerning the efficiency of random copolymers, we conclude that these substances are active exclusively if they are only sparingly soluble in both components of the blend.

ACKNOWLEDGEMENT

We are grateful to the "Materials Science Research Centre of the University Mainz" for financial support.

REFERENCES

1. S. Enders, A. Huber, B.A. Wolf, *Polymer* **35**, 5743 (1994)
2. S. Enders, B.A. Wolf, K. Binder, *J. Chem. Phys.* **103**, 3809 (1995)
3. M. Heinrich, B.A. Wolf, *Polymer* **33**, 1926 (1992)
4. Z. Tong, K. Meissner, B.A. Wolf, *Macromol. Chem. Phys.* **196**, 521 (1995)
5. M. Heinrich, B.A. Wolf, *Macromolecules* **25**, 3817 (1992)
6. A. Stammer, B.A. Wolf, *Polym. Commun.*, submitted.
7. A. Stammer, Ph.D. Thesis, Mainz 1997
8. M. Wagner, B.A. Wolf, *Macromolecules* **26**, 6498 (1993)
9. A. Schneider, B. A. Wolf, *Macromol. Rapid Commun.* **18**, 561 (1997)
10. M. Wagner, B.A. Wolf, *Polymer* **34**, 1460 (1993)
11. U. Jorzik, B.A. Wolf, *Macromolecules* in press
12. I. Welge, M.S. Thesis, Mainz 1997
13. A. Stammer, B.A. Wolf, *Macromol. Rapid Commun.* **19**, 123 (1998)

Silane Treatment in Polypropylene Composites: Adsorption and Coupling

Béla Pukánszky* and Zoltán Demjén

Technical University of Budapest, Department of Plastics and Rubber Technology
H-1521 Budapest, P.O. Box 92, Hungary
Central Research Institute for Chemistry, Hungarian Academy of Sciences
H-1525 Budapest, P.O. Box 17, Hungary

Abstract: The study of the effect of eight silane coupling agents on mechanical properties of PP/$CaCO_3$ composites showed that the application of certain compounds leads to the reactive coupling of the components. The adsorption of the silanes on the surface of the filler was studied by a dissolution method, while the structure of the adsorbed layer was determined by FTIR spectroscopy and GPC measurements. Model reactions and an FTIR study were carried out to reveal possible coupling reactions of PP and the active silanes. The results showed that amino-functional silane coupling agents adsorb on the surface of the filler, polycondensation taking place and a polysiloxane layer of high molecular weight forms as a result. The model reactions indicated that during processing, PP oxidizes even in the presence of stabilizers and oxygen-containing functional groups form on the polymer chains. These easily react with the amino functionality of the silane coupling agents strongly bonding the polymer to the treated filler.

INTRODUCTION

Organofunctional silanes are extensively used for the surface treatment of glass fibers and glass beads introduced into thermoset matrices (Refs. 1,2). These reactive coupling agents form chemical bonds with the active hydroxy groups of the filler and their functional group reacts with the resin during curing (Ref. 3). The strong covalent bonds formed in these reactions make possible efficient stress transfer and the utilization of the reinforcing effect of fibers. The success of these coupling agents in thermoset composites leads to extensive attempts to use them also in other polymers in spite of the fact that the proper conditions of reactive coupling do not always exist. Polypropylene is a nonpolar polymer without any reactive groups on the chain; thus, silanes can hardly be expected to act as efficient coupling agents in its composites. Nevertheless, occasionally reactive coupling and improvement of mechanical properties were observed even in particle-filled PP composites (Ref. 4), but an acceptable explanation has not been given for the phenomenon so far. Another difficulty is expected in PP/$CaCO_3$ composites since this filler does not contain active OH groups on its surface.

 CCC 1022-1360/99/$ 17.50+.50/0

In an extended study (Refs. 5-8), a $CaCO_3$ filler was surface-treated with eight organofunctional silanes, both commercial and laboratory products, composites were prepared and their tensile properties determined. The efficiency of the silanes as coupling agents was checked and quantitatively evaluated by a simple semiempirical model developed earlier (Refs 9,10). In some cases, reactive coupling was observed in accordance with literature references (Ref. 4). This effect is well demonstrated by Fig. 1 where tensile strength of the composites is plotted against the surface coverage of the filler. It is apparent that the aminosilane has a strong reactive coupling effect, while stearic acid, used as a reference, and the other silane coupling agents act as nonreactive treating agents, surfactants. These apparently contradictory results initiated a detailed study of the interactions between the filler and coupling agents on the one hand, and between the silanes and the polymer on the other. This paper contains the most important results of this study and attempts to give a tentative explanation both for the different behaviour of the coupling agents and the mechanism of reactive coupling in PP/$CaCO_3$ composites.

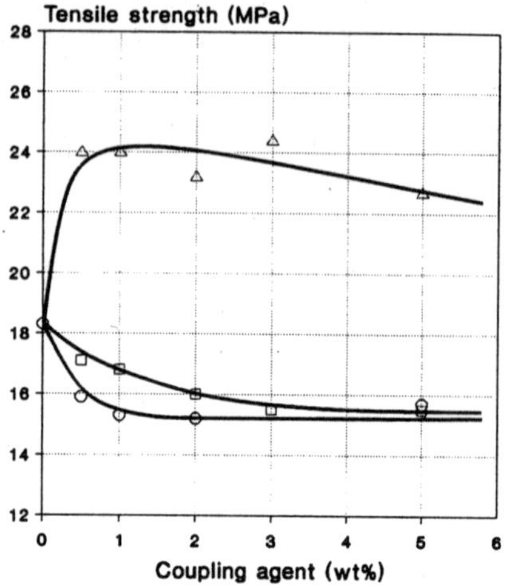

Fig. 1 *Effect of silane coupling agents on the strength of PP/$CaCO_3$ composites; (Δ) AMPTES, (□) SPTES, (O) STAC (for codes, see Table 1); filler content: 20 vol.-%*

EXPERIMENTAL

The $CaCO_3$ filler used (Omyacarb 2GU, Omya, Switzerland) had an average diameter of 2.5 μm and specific surface area of 3.59 m^2/g. The silane coupling agents used are listed in Table 1. The chemical structure of the original and adsorbed silanes was characterized by FTIR spectroscopy. A Mattson Galaxy 3020 apparatus was used for the registration of transmittance spectra, while DRIFT spectra were taken by a Bio-Rad FTS 65A/896 apparatus.

Surface treatment was carried out by two different techniques, dry blending and adsorption from butan-1-ol solution. The first procedure was used for the determination of the adsorption of the silane on the filler surface (dissolution method; Refs 11,12). 200 g of $CaCO_3$ was

Table 1 Coupling agents used for surface treatment

Coupling agent	Formula	Grade	Source
SPTES	$CH_3(CH_2)_{16}COO(CH_2)_3Si(OCH_2CH_3)_3$	–	laboratory
PPTES	$CH_3(CH_2)_7COO(CH_2)_3Si(OCH_2CH_3)_3$	–	laboratory
ACPTES	$CH_3COO(CH_2)_3Si(OCH_2CH_3)_3$	–	laboratory
MPTMS	$CH_2{=}C(CH_3)COO(CH_2)_3Si(OCH_3)_3$	GF 31	Wacker
CVBS	$CH_2{=}CHC_6H_4CH_2NH(CH_2)_2NH(CH_2)_3Si(OCH_3)_3 . HCl$	Z 6032	Dow Corning
GPTES	$\underbrace{CH_2\text{-}CH}_{O}\ CH_2O(CH_2)_3Si(OCH_2CH_3)_3$	SLM 50-882	Wacker
CHAPTMS	cyclohexyl-$NH(CH_2)_3Si(OCH_3)_3$	SLM 50-892	Wacker
AMPTES	$NH_2(CH_2)_3Si(OCH_2CH_3)_3$	GF 93	Wacker
STAC	$CH_3(CH_2)_{16}COOH$	–	–

dry-blended with various amounts of the silane coupling agents (0-5 wt.-%) and the filler was stored for two days under ambient conditions. 20 g of the treated filler was stirred in 100 ml of cyclohexane or methanol for 30 min and then left standing for a day. The suspension was centrifuged and the silane content of the solution was determined by FTIR or HPLC. The amount of adsorbed silane was calculated from the measured concentrations. The adsorption of aminosilanes was determined in methanol, while that of all other compounds in cyclohexane. In the other procedure, 700 g of $CaCO_3$ was stirred in 700 ml of butan-1-ol for 3 h, the slurry was left standing for 3 days and then the solvent was removed under vacuum. Prehydrolysis, usually recommended in the literature (Ref. 3), was not used in either procedure.

The structure of the adsorbed silane was characterized by DRIFT and FTIR before and after washing with THF. The molecular weight of the dissolved silanes was determined by GPC (Waters 510, Viscotek 200 detector).

Reactions between PP (Tipplen H 327, TVK, Hungary; MFI 10 g/10 min at 230 °C and 21.6 N) and the silanes were carried out in an internal mixer (Brabender W 50 EH) at 190 °C. 2 ml of a silane coupling agent was added to the melt of 35.9 g PP. Approximately 0.1-g samples were taken in 5- or 10-min intervals from the mixture. They were compression-moulded (Fontijne SRA 100) into 100-μm thin films, then extracted either in chloroform or methanol depending on the silane coupling agent used in the reaction. Extraction was carried out for 12 h in a Soxhlet extractor. FTIR spectra of the films were recorded before and after extraction. Various model reactions were also carried out. Oxidation of PP was accelerated by the introduction of O_2, while cumene hydroperoxide was added to increase the concentration of free radicals in the melt. The reactions were carried out both with neat PP and also in the presence of silanes.

RESULTS AND DISCUSSION

Principal condition of reactive coupling is chemical reaction, or at least a very strong interaction of both components with the coupling agent. As a consequence, we must study first the filler/coupling agent, then the coupling agent/polymer interaction in order to find a plausible explanation for the observed reactive coupling effect in the non-reactive system studied. The results of these investigations are reported in the following paragraphs.

Adsorption of coupling agents

As mentioned before, the interaction of the filler and the polymer can be characterized by a dissolution technique developed earlier (Refs. 11,12). Dissolution curve characteristics of nonreactive coupling are presented in Fig. 2. The amount of the bonded coupling agent is plotted against the coupling agent concentration used for the treatment. At low amounts, all the coupling agent is bonded to the surface, but at higher concentrations, equilibrium adsorption is reached. Two characteristic quantities can be derived from the dissolution curve: the concentration of the proportionally bonded coupling agent (c_{100}) and the value of maximum adsorption (c_{max}). c_{100} approximately corresponds to the monolayer coverage of the filler by the coupling agent and its value depends on the specific surface area of the filler, on the surface requirement and orientation of the low-molecular-weight compound, and on other conditions of

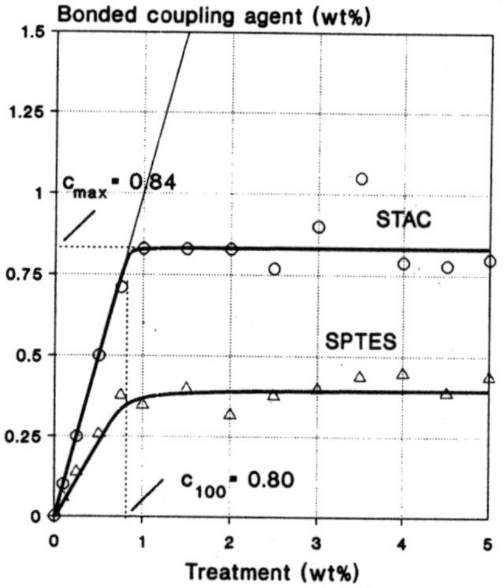

Fig. 2 *Dissolution curves of STAC and SPTES*

the dissolution measurement (solvent, temperature, concentration). According to the figure, the silane containing a long alkyl chain (SPTES) behaves very similarly to stearic acid, the surfactant most often used for the non-reactive treatment of $CaCO_3$. The only difference is found in the first, proportionally bonded region of the dissolution curve, the slope being here smaller than unity. The decreased slope is probably the result of the complex competitive processes of adsorption and dissolution taking place during the measurement (adsorption of silane and solvent, dissolution of silane). Since PPTES has a structure similar to SPTES, its dissolution is also very similar to the one presented in Fig. 2.

The type and size of the organofunctional group play an important role in the adsorption of the coupling agent on the filler surface. As Fig. 3 shows, MPTMS adsorbs in larger amounts and the adsorption continues up to the largest concentration used. A further increase in the amount of adsorbed coupling agent is shown by amino-functional silanes, virtually the total amount of AMPTES being bonded to the filler surface. These results prove that silane coupling agents adsorb on the surface of $CaCO_3$. The amount and strength of the adsorption depend on the structure of the silane and primarily on the size and character of the organofunctional group. Aminosilanes bond especially strongly to the surface, probably due to the catalytic effect of the amino group which increases the rate of polycondensation of the silane.

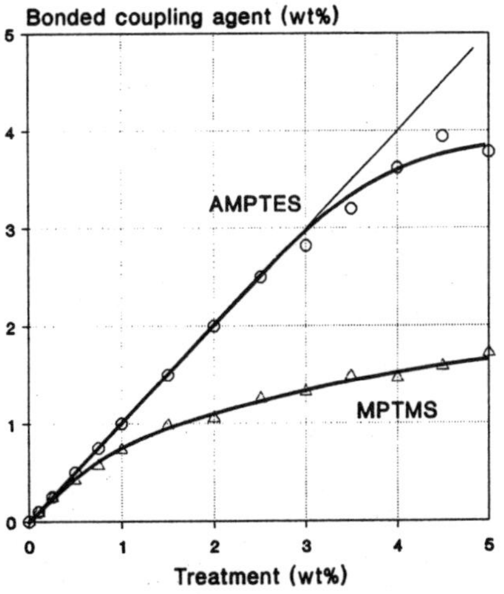

Fig. 3 *Dissolution curves of MPTMS and AMPTES*

The FTIR spectrum of SPTES and the DRIFT spectrum of $CaCO_3$ treated with this coupling agent are presented in Fig. 4. The figure clearly shows that profound changes take place in the structure of the silane due to adsorption. Analysis of the spectra and comparison with literature data show that polycondensation of the silane takes place during treatment and the following storage. The spectrum indicates that a polysiloxane with cage-like polycyclic structure forms

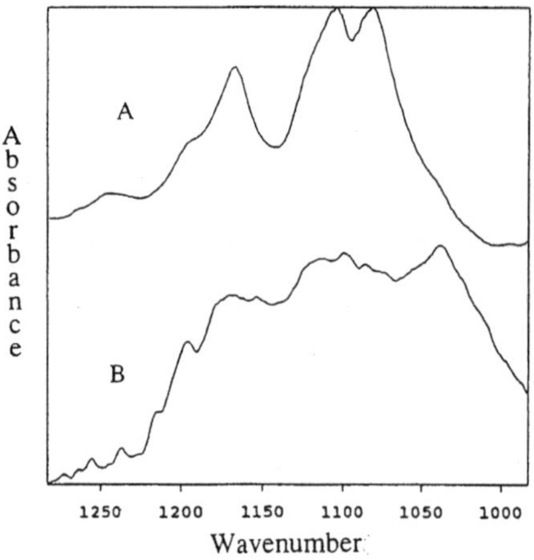

Fig. 4 *(A) FTIR spectrum of neat SPTES, (B) DRIFT spectrum of adsorbed SPTES*

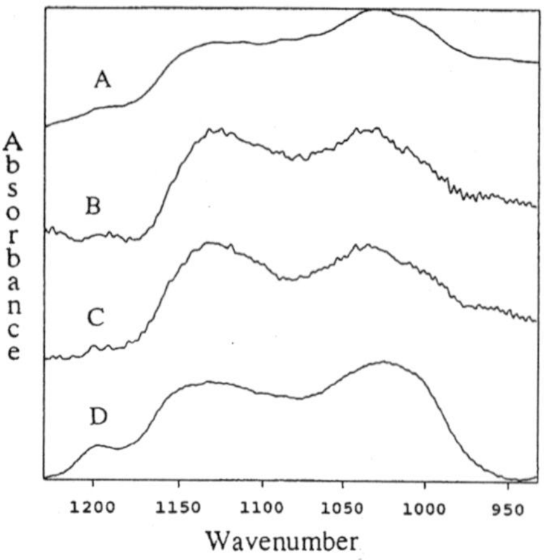

Fig. 5 *Spectra of polysiloxane formed from AMPTES: (A) butan-1-ol treatment, (B) dry-blending, (C) after washing with THF, (D) neat AMPTES film*

from SPTES (Refs. 13,14). This result verifies our procedure and proves that prehydrolysis of the silane before treatment is not necessary. The cage-like structure leads to poor resistance to solvents; the polysiloxane can be completely removed from the surface by washing with THF. Also AMPTES forms a polysiloxane layer on the surface of the filler, but the structure and properties of the polymer are completely different. Parts of four spectra are compared in Fig. 5. Spectrum A was obtained on the filler treated with butan-1-ol. Virtually the same spectrum is obtained when the treatment was carried out by dry blending (B). The spectrum does not change on washing the filler with THF (C) and completely corresponds to that of a polysiloxane film prepared by polycondensation of neat AMPTES (D). The spectrum of the hard, transparent film obtained by simple storage of AMPTES in a crystallization dish under laboratory conditions was recorded by transmittance FTIR. The results unambiguously prove several points: (i) Polymerization of the silane takes place on the surface of the filler. (ii) The structure of the polysiloxane is the same in all cases. (iii) The treatment technology does not influence the structure of the adsorbed layer. (iv) The structure of the film formed from AMPTES is different from that of SPTES, an open, ladder-type polysiloxane network being formed in the former case. (v) The structure of the adsorbed layer depends mainly on the size and type of the organofunctional group of the silane.

Table 2 Fraction of bonded silane, relative amount and molecular weight of the polysiloxane fractions dissolved from the filler surface by THF

Coupling agent	Bonded silane (%)	Molecular weight					
		1 000 - 4 000		4 000 - 50 000		> 50 000	
		M_w	amount	M_w	amount	M_w	amount
SPTES	0	3 080	1.00	4 140	1.04	131 300	0.08
PPTES	12	2 330	1.00	5 070	2.30	166 170	0.40
ACPTES	43	1 290	1.00	5 580	2.90	169 400	3.70
MPTMS	0	–	–	–	–	–	–
GPTES	89	1 660	1.00	–	–	149 000	0.80
CVBS	45	–	–	37 400	1.00	142 800	3.10
CHAPTMS	56	–	–	–	–	136 900	1.00
AMPTES	93	–	–	–	–	139 500	1.00

Both the molecular weight of the polysiloxane layer and its resistance to THF depend on the chemical structure of the silane. Molecular weights of different fractions and the amount of the adsorbed polymer determined from the FTIR spectra after washing with THF are presented in Table 2. The molecular weight distribution of the polysiloxane was usually bimodal, sometimes even more than two peaks being observed (Fig. 6). The heights of the peaks were always related to the smallest one, which gives an estimate of the relative amount of different fractions. As Table 2 and Fig. 6 show, the molecular weight and the amount dissolved change strongly from one silane to the other. The silanes with long chains give polymers of low molecular weights which can be dissolved from the surface of the filler with THF easier than those formed from silanes with smaller organofunctional groups. Aminosilanes invariably transform to high-molecular-

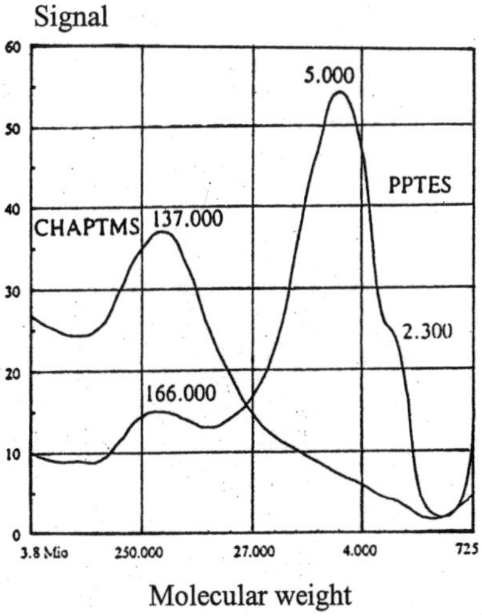

Fig. 6 *Molecular weight distribution of polysiloxanes dissolved from the surface of $CaCO_3$ treated with PPTES and CHAPTMS*

weight polysiloxanes, which strongly resist dissolution with THF.

The results clearly prove that polycondensation of silane coupling agents occurs always on the surface of an inorganic filler and, in spite of the absence of chemical coupling, the polymer formed may be attached strongly to the surface. The possibility of a strong coupling agent/filler interaction having thus been proved, the chemical coupling of the polysiloxane layer and polypropylene must be now verified.

Coupling reactions

A great number of papers were published on the thermooxidative degradation of polymers including PP (e.g., Refs. 15,16). During the processing of the polymer under atmospheric conditions, hydroperoxides form and β-scission of the polymer takes place leading to the formation of oxygen-containing groups on the chain (Ref. 16). Rabek (Ref. 15) analyzed the reaction products and determined their relative amounts under various processing conditions showing that a considerable amount (~25 %) of carboxylic acids forms in the polymer.

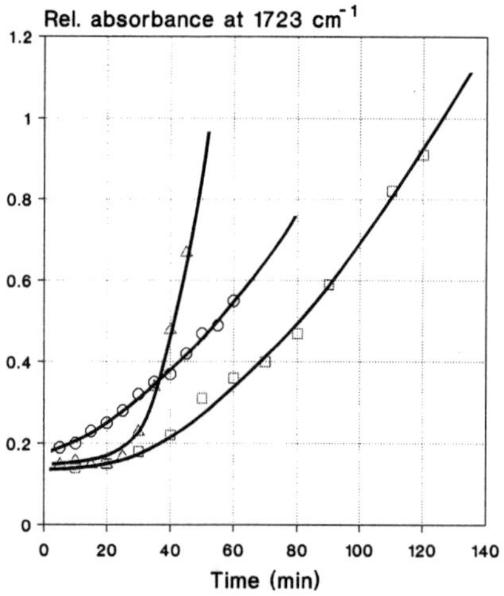

Fig. 7 *Carbonyl formation in PP as a function of homogenization time at 190 °C, (□) neat PP, (O) cumene hydroperoxide added, (Δ) oxygen presence*

Model reactions were carried out in order to check the possibility of carbonyl formation under the condition of sample preparation. Carbonyl groups were detected in the PP matrix used for the preparation of our composites in spite of the presence of stabilizers (Fig. 7). The new absorption band detected at 1723 cm^{-1} was assigned to the carbonyl group of saturated ketones, aldehydes and acids (Ref. 17). The presence of cumene hydroperoxide, i.e. an increase in the concentration of free radicals, accelerates the formation of carbonyl groups as it is shown in Fig. 7. Oxygen exerts its effect after 20-30 min of mixing, when the carbonyl group content in the polymer sharply increases. Although the exact chemical reactions leading to these changes are not known, it is clear that functional groups form during the processing of PP even in the presence of stabilizers. One might object that the reaction times used here are long compared with the time of homogenization. However, we must bear in mind that the sensitivity of IR spectroscopy is not exceptionally high and that a few bonds are sufficient to increase interaction significantly. According to the results, the carbonyl groups and in particular the acid groups are the only candidates which can react with the coupling agents forming covalent bonds.

Possible reactions of three silane coupling agents were studied in further model reactions. Because of limited space, only the results obtained on AMPTES are presented here. Extraction of the films used for the monitoring of chemical changes in the polymer gave unambiguous proof of the coupling of the silane and the polymer. Only a limited extent of such coupling occurred between PP and MPTMS, but AMPTES was already totally bonded to the polymer after a short mixing time. The spectra of the samples taken after 10 min of mixing are virtually identical before and after extraction (Fig. 8, spectra 2 and 3). The same conclusion can be drawn from Fig. 9 showing the changes in the concentration of various functional groups as a function of mixing time.

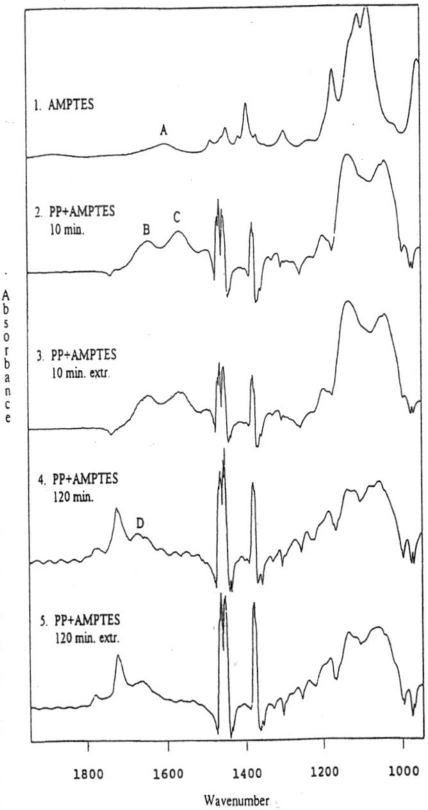

Fig. 8 *FTIR difference spectra of the reaction product of PP and AMPTES*

The absorbance at 1596 cm^{-1} appearing in the spectrum of AMPTES (Fig. 8, spectrum 1, A) can be assigned to the deformational vibration of the N-H bond in the primary amine group (Ref. 17). After 10-min reaction time, this band disappears and two bands appear at lower and higher wavenumbers, at 1640 (B) and 1563 cm^{-1} (C). According to Refs. 18-21, the amide I (C=O) vibration of secondary amides in the solid state appears between 1680-1630 cm^{-1}.

The intensity of deformational vibration of the N-H bond of a secondary amide absorbs in the range of 1581-1518 cm^{-1} (see Ref. 17). Accordingly, we may assume that the above mentioned vibrations at 1640 and 1563 cm^{-1} are the amide I and amide II bands of a secondary amide, respectively. We showed above (see Fig. 7) that significant amount of carboxyl groups form during the processing of PP, which can react with the amino functionality of AMPTES.

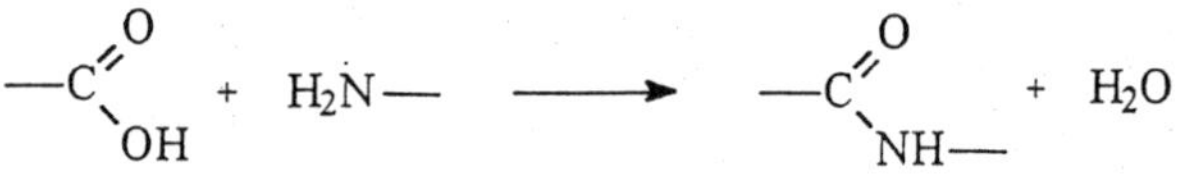

(1)

The intensity of the deformational vibration of the N-H bond (amide II) gradually decreases with increasing reaction time, while the amide I absorption shifts to higher wavenumbers. After 120 min, an absorption band appears at 1670 cm^{-1} (Fig. 8, Spectrum 4, D). The double band appearing at 1790-1720 cm^{-1} and 1710-1670 cm^{-1} can be assigned to the symmetric and antisymmetric C=O vibration of imides (Refs. 17,22,23). This suggests the possibility that the secondary amide enters into reaction with another acid group forming imides:

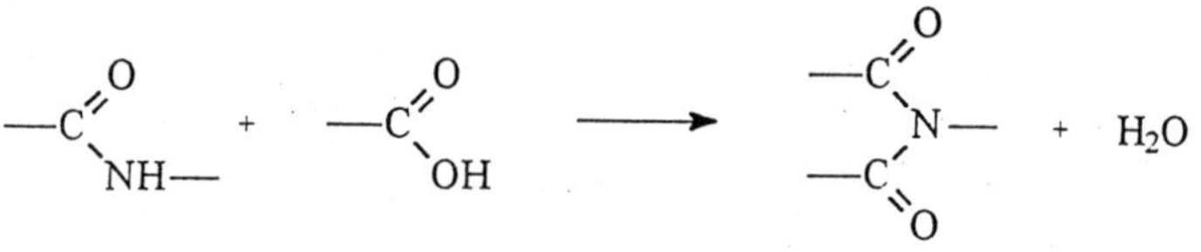

(2)

Theoretically, the tautomeric conversion of the amide to unstable imidic acid is also possible, which may react further with the free, not hydrolyzed -Si-O-Et groups resulting in the stable silyl ester of imidic acid:

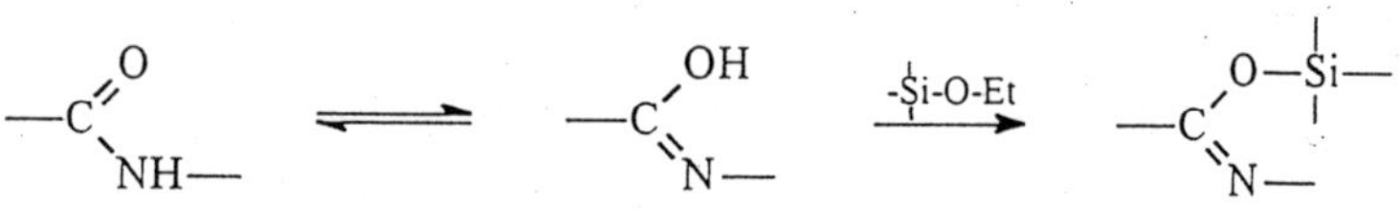

(3)

The C=N vibration of the formed imidate appears at 1690-1645 cm^{-1} (see Ref. 23). This may explain the appearance of band D around 1670 cm^{-1} and the disappearance of the amide II band as well. However, the existence of free ethoxysilyl groups in the melt at long reaction times is not probable.

Time dependence of the changes in the concentration of various functional groups is presented in Fig. 9. The intensity of carbonyl absorbance increases with reaction time (P in Fig. 9) similarly to the virgin PP (Fig. 7). However, in the AMPTES/PP reaction, the carbonyl content increases only after an induction period of 30-40 min, its value up to this point being zero. The difference may be explained by the immediate reaction of the primary amino group of AMPTES with the forming carboxyl groups of the polymer. After approximately 40 min, all free NH_2 groups are consumed and carbonyl absorption appears in the spectrum. After a slight initial decrease, the total amount of amide and imide (or imine) groups (B, D) seems to remain constant up until the end of the reaction. The amide groups transform into imides according to (2) resulting in a continuous decrease in the NH content (Fig. 9, C). After 80 min, all the secondary amide is consumed, the total amount having been converted to imide (or imidate).

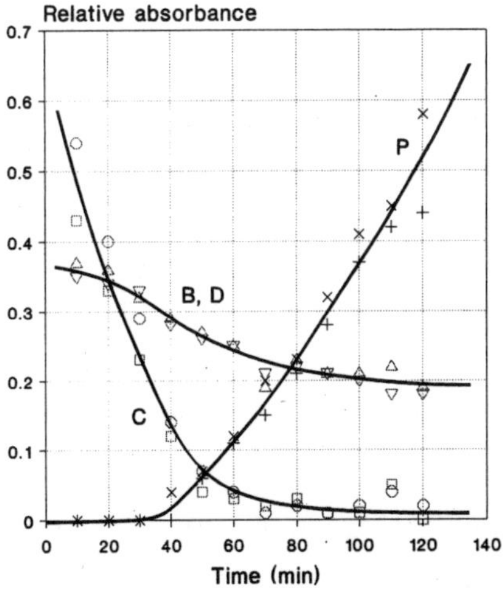

Fig. 9 *Reaction of PP with AMPTES, change in functional group content with reaction time; P: carbonyl, C: amide II, B,D: amide I and imide, (x, O, Δ) before extraction, (+, □, ▿) after extraction*

After 10-min reaction time, the anti symmetric stretching vibration of the triethoxysilyl group of AMPTES at 1103 and 1080 cm^{-1} completely disappears and a new double band appears at 1128 and 1039 cm^{-1}. These absorption bands are characteristic of high-molecular-weight polysiloxane chains (Refs. 13,14) (Fig. 8, spectrum 2). According to the spectrum, AMPTES reacts with some moisture present (see, e.g., reactions 1 and 2), hydrolysing and condensing to polysiloxane even at a high reaction temperature (190 °C).

The results show that AMPTES enters into chemical reactions with the polymer matrix immediately after melting. The chemical bonds formed are stable, they cannot be disrupted even by 12-h extraction. FTIR spectra indicate that secondary amide forms in the chemical reaction between the primary amino group of AMPTES and the carboxyl groups continuously forming in the PP. Since this reaction seems to proceed very fast, we can assume that it takes place also during the homogenization of PP/$CaCO_3$ composites.

CONCLUSIONS

The above presented results clearly show that reactive coupling of the components is possible in a relatively inert system like the PP/$CaCO_3$ composite if organofunctional silane coupling agents are used. Surface treatment of the filler with various silanes always leads to the polycondensation of the coupling agent. Polymerization takes place even in the absence of prehydrolysis and the structure of the polysiloxane layer is independent of the surface treatment technique. On the other hand, both the structure and molecular weight of the polysiloxane depend very much on the organofunctional group of the coupling agent. Large groups lead to a low- molecular-weight polymers, which can be easily dissolved from the surface of the filler. Small groups result in a high-molecular-weight product, which adheres more strongly to the filler. Polycondensation of aminosilanes is especially fast and also adhesion of the polymer is very strong, which might be explained by the catalytic effect of the amino group on the condensation reaction.

Model reactions proved that PP is oxidized during processing even in the presence of stabilizers leading to the formation of reactive groups on the polymer chain. These groups react with the amino functionality of the silane forming strong amide groups. The strong adhesion of the polysiloxane layer on the surface of the filler and the chemical reaction of this polymer layer with PP leads to the observed reactive coupling of the otherwise non-reactive components, PP and $CaCO_3$.

ACKNOWLEDGEMENTS

The authors wish to express their gratitude to Dr. M. Szesztay, Central Research Institute of Chemistry, Hungarian Academy of Sciences for her help in GPC analysis and to Dr. Ottó Berkesi, József Attila University of Szeged, for the assistance in the DRIFT measurements. The authors are also grateful for the financial support of the National Scientific Research Fund of Hungary (Grant No. T 016500).

REFERENCES

(1) H. Ishida, *Interfaces in Polymer, Ceramic and Metal Matrix Composites*, Elsevier, New York, 1988

(2) E.P. Plueddemann, *Silane Coupling Agents*, Plenum, New York, 1982

(3) H. Ishida, J.D. Miller, *Macromolecules* **17**, 1659 (1984)

(4) J.P. Trotignon, J. Verdu, R. Boissard, A. Vallois, in *Polymer Composites* (B. Sedláček, Ed.), p. 191, Walter de Gruyter, Berlin, 1986

(5) Z. Demjén, B. Pukánszky, *Polym. Compos.* **18**, 741 (1997)

(6) Z. Demjén, B. Pukánszky, J. Nagy, *Composites* 29**A**, 323 (1998)

(7) Z. Demjén, B. Pukánszky, E. Földes, J. Nagy, *J. Colloid Interface Sci.* **190**, 427 (1997)

(8) Z. Demjén, B. Pukánszky, J. Nagy Jr., *Polymer* **40**, 1763 (1998)

(9) B. Pukánszky, B. Turcsányi, F. Tüdős, in *Interfaces in Polymer, Ceramic, and Metal Matrix Composites* (H. Ishida, Ed.), p. 467, Elsevier, New York, 1988

(10) B. Pukánszky, *Composites* **21**, 255 (1990)

(11) E. Papirer, J. Schulz, C. Turchi, *Eur. Polym. J.* **20**, 1155 (1984)

(12) E. Fekete, B. Pukánszky, A. Tóth, I. Bertóti, *J. Colloid Interface Sci.* **135**, 200 (1990)

(13) H. Ishida, D.J. Miller, *J. Polym. Sci., Polym. Phys. Ed.* **23**, 2227 (1985)

(14) J.F. Brown Jr., L.H. Vogt Jr., P.I. Prescott, *J. Am. Chem. Soc.* **86**, 1120 (1964)

(15) J.F. Rabek, in *Degradation of Polymers* (C.H. Bamford, C.H.F. Tipper, Eds.), Chap. 4, p. 425, Elsevier, Amsterdam, 1975

(16) S.S. Stivala, J. Kimura, S.M. Gabbay, in *Degradation and Stabilization of Polyolefins*, (N.S. Allen, Ed.), p. 63, Appl. Sci. Publ., London, 1983

(17) G. Socrates, *Infrared Characteristic Group Frequencies*, J.Wiley, Chichester, 1980

(18) R.A. Nyquist, *Spectrochim. Acta* **19**, 509, 713, 1595 (1963)

(19) R.D. Mclachlan, R.A. Nyquist, *Spectrochim. Acta* **20**, 1397 (1964)

(20) M. Beer, *J. Chem. Phys.* **29**, 1097 (1958)

(21) Y. Kuroda, *Spectrochim. Acta, Part A* **29**, 411 (1973)

(22) T. Uno, K. Machida, *Bull. Chem. Soc. Jpn.* **34**, 545, 551 (1961)

(23) S. Holly, P. Sohár, *Infravörös spektroszkópia*, Műszaki Könyvkiadó, Budapest, 1968

Plasma Deposition of Silicon-containing Layers on Polymer Substrates

Fabio Garbassi*[1], Ernesto Occhiello[2]

[1] EniChem S.p.A., Centro Ricerche Novara "Istituto Guido Donegani", Via Fauser, 4, 28100 Novara, Italy
[2] EniChem S.p.A., Div. Poliuretani e Cloro, Via Taramelli, 26, 20124 Milano, Italy

Abstract: Plasma-polymerized coatings have been deposited on polymer substrates using mixtures of silicon-containing monomers with O_2 and/or CF_4. With ternary gas mixtures, novel silicon oxyfluoride coatings were obtained. By oxygen plasma post-treatment, improvements in barrier properties of polymer substrates to gases were observed.

INTRODUCTION

Thin films deposited from plasma polymerization exhibit very interesting and sometimes unique properties (Ref. 1). On metal substrates, the preferred applications range from protective coatings to biocompatibility. On polymer substrates, the most interesting issues were barrier, biocompatibility, wear and friction, though other applications have been pursued, like membrane permselectivity, electrical insulation, moisture sensors and UV protection (Ref. 2 and references therein).

For improving barrier properties of polymer films against small molecules like oxygen, water and carbon dioxide, deposition of thin inorganic coatings was found advantageous mainly in food packaging applications (Ref. 3). While aluminium deposition is the preferred solution on industrial scale, the deposition of silica-like coatings was found interesting when transparent layers are preferred. Provided that the deposited coating is dense, free of defects and resistant to mechanical load (necessary for surviving in further processing, like lamination of a sealable film), an improvement of barrier properties comparable to that of obtained by aluminium metallization can be achieved (Refs. 3,4). However, the economics of the process is not yet well assessed because it is necessary to start from costly SiO, instead of SiO_2, which needs a very high evaporation temperature. Furthermore, whereas SiO, when evaporated in the presence of oxygen, gives rise to a dense and glassy SiO_x ($x = 1.5$-1.7) coating, SiO_2 forms a nanoporous layer permeable to small gas molecules (Ref. 4).

An alternative solution is the plasma deposition of siloxane monomers like hexamethyldisiloxane (HMDSO) (Ref. 5) and tetramethyldisiloxane (TMDSO) (Ref. 6). It was

 CCC 1022-1360/99/$ 17.50+.50/0

found that, in order to obtain a barrier improvement, some oxygen must be present in the plasma discharge atmosphere. In this way, the nature of polysiloxane deposited by plasma can be shifted from organic (silicone-like) to inorganic (silica-like) character by controlling the plasma parameters (Ref. 7). This work concerns the plasma deposition from siloxane monomers on polyolefin substrates, using some original experimental procedures like the presence of fluorinated monomers in the discharge gas feed, or post-deposition plasma treatment of polysiloxane coating.

RESULTS AND DISCUSSION

Plasma polymerization

A parallel-plate plasma reactor was used for deposition experiments. As substrates, several polymer plates or films have been used, made of high-density polyethylene, HDPE (1mm thick), linear low-density polyethylene, LLDPE (0.02 mm thick), biaxially oriented poly(ethylene terephthalate), PET (0.012 mm thick) and polyamide 6, PA-6 (0.012 mm thick). Hexamethyldisiloxane (HMDSO) and tetramethyldisiloxane (TMDSO), used as polymerizable monomers, were introduced into a glass vessel connected to the plasma reactor through a glass valve and a glass-metal joint. The temperature of the vessel was maintained at 30 °C. They were used alone or in the presence of other gases, like oxygen, CF_4 or their mixtures.

The deposition rate was monitored by a quartz crystal microbalance (Intellemetrics).

Characterization

Chemical composition of deposited layers was determined by X-ray photoelectron spectroscopy (XPS), using a PHI model 5500 XPS spectrometer. Layer thickness was measured by depth profiling carried out with combined XPS and ion bombardment. A sputter rate of 1.5 nm/min was adopted, as calibrated with a Ta_2O_5/Ta standard specimen.

Wettability was determined by contact angle measurements, using the sessile drop method and a Ramé-Hart goniometer. Each contact angle value is the average of at least five different measurements on the same sample sheet.

Permeability to oxygen was measured by a Lissy GPM-200 gas permeability fractometer. This instrument measures isostatic permeability, i.e., without substantial pressure differences on the two sides of the sample.

More details on experimental procedures can be found elsewhere (Ref.8).

Results

In order to determine the effect of experimental conditions and to optimize the application performance of deposited films, the following parameters were taken into account:

- nature of monomers or monomer mixtures, and their composition;
- plasma polymerization parameters, i.e., power, gas flow rate, time;
- nature of the substrate.

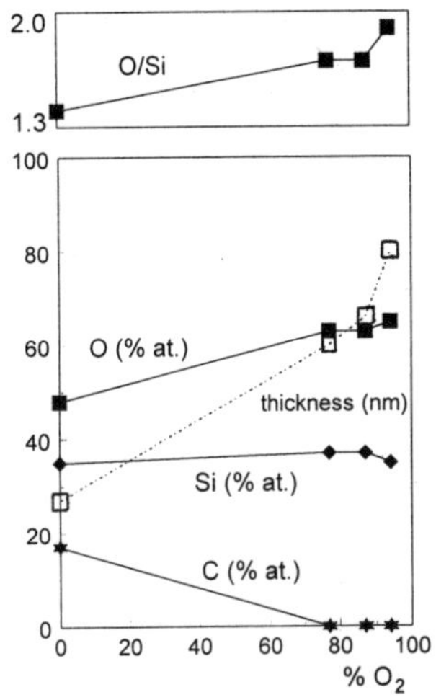

Fig. 1 Surface composition and thickness of plasma-deposited coatings from HMDSO/O_2 mixtures vs. O_2 amount.

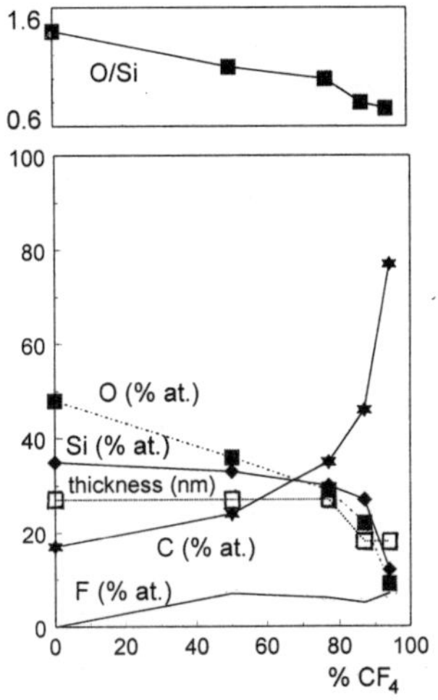

Fig. 2 Surface composition and thickness of plasma-deposited coatings from HMDSO/ CF_4 mixtures vs. CF_4 amount.

The results are presented in Figs. 1-9. Figure 1 shows the effect of the presence of increasing amounts of O_2 in the HMDSO plasma. The deposition rate, as suggested by the increased thickness of the coating, increases, while the chemical character of the surface is converted from partially organic to inorganic, as suggested by the complete elimination of carbon. The O/Si atomic ratio slightly increases with the increase in oxygen amount in the gas feed, reaching at the highest O_2 concentration a value near that expected for SiO_2, viz. 2. Such results are in agreement with previous literature (Refs. 9,10) and suggest a better efficiency of oxygen-rich mixtures in abstracting the organic portion of siloxane. Figure 2 reports the effect of an increasing amount of CF_4 in the HMDSO plasma. In this case, where CF_x radicals are formed in the plasma (Refs. 11,12), a coating containing a high amount of carbon is deposited. The carbon content increases with the CF_4 concentration in the gas feed. The coating thickness is always low and tends to decrease, due to the high probability of formation of volatile compounds. Some fluorine is introduced in the coating, thanks to the reaction of $\dot{C}F_x$ species with the polysiloxane chains. The O/Si ratio is always low, assuming a value around unity and showing some fair tendency to decrease.

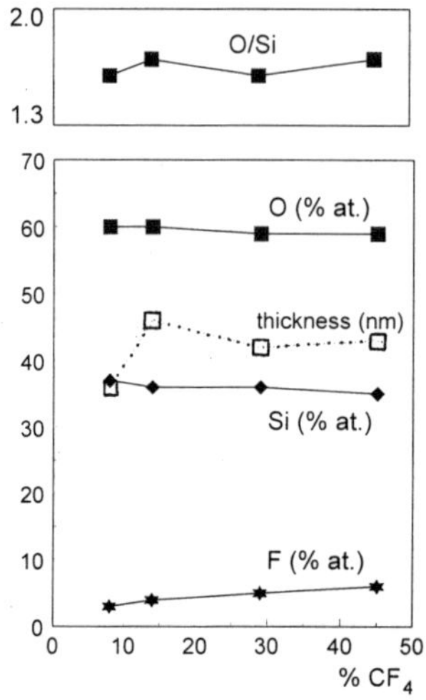

Fig. 3 Surface composition and thickness of plasma-deposited coatings from HMDSO/ O_2/CF_4 mixtures vs. CF_4 amount (HMDSO/ O_2 = 1/5).

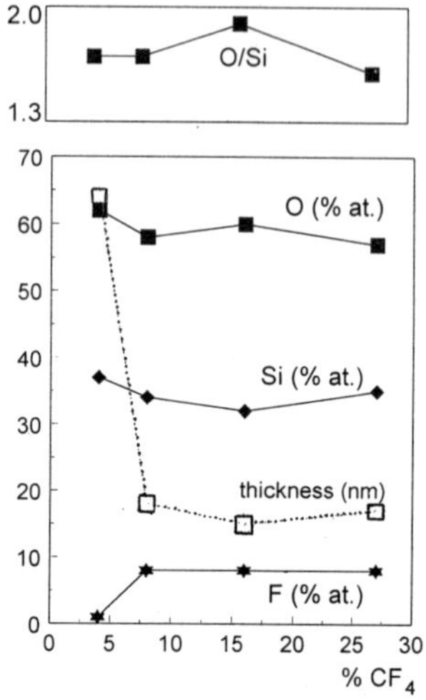

Fig. 4 Surface composition and thickness of plasma-deposited coatings from HMDSO/ O_2/CF_4 mixtures vs. CF_4 amount (HMDSO/ O_2 = 1/12.5).

In Figs. 3 and 4, the effects on coating properties of the presence of O_2/CF_4 gas mixtures at two different oxygen concentration levels are presented. Gas feed compositions were chosen in order to maintain always a relatively high oxygen concentration. In these conditions, in addition to abstracting the organic moiety from the coating as described above, also the formation of atomic fluorine in the discharge is favoured, as described in the literature (Refs. 11,12). In both cases, a silicon oxyfluoride coating not containing carbon residues is clearly formed, having O/Si atomic ratios around 1.7. F/Si ratios behave differently changing the oxygen concentration: at the lower level, F/Si shows a fair tendency to increase (from 0.08 to 0.17) with increasing CF_4, while at the higher level the ratio assumes a fairly constant value around 0.24, apart from an exception when the the O_2/CF_4 is very high (25). In the latter case, a F/Si of 0.03 was measured. Because in the same conditions, a rather high thickness of the deposited coating was observed relative to others, it is likely that in such conditions lower amounts of atomic fluorine are formed in the plasma. In the other cases, the low amounts of fluorine in the coating (relative to oxygen) is related both to the tendency of forming volatile compounds and the difficulty to accommodate fluorine atoms in the silica-like coating structure.

As expected, the thickness of coatings derived from ternary mixtures is always lower than that from binary HMDSO/O_2 mixtures because the presence of CF_4 leads to a higher tendency to etching and forming volatile compounds.

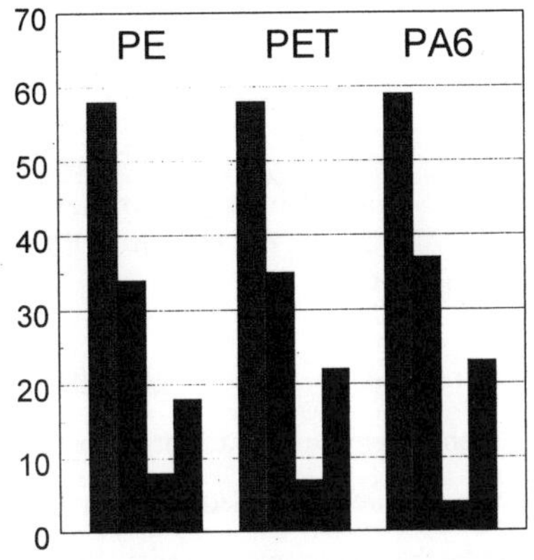

Fig. 5 Surface composition (% at.) and thickness (nm) of plasma-deposited coatings from HMDSO/O_2 /CF_4 mixtures on different substrates. Bars represent from left to right O, Si, F and thickness.

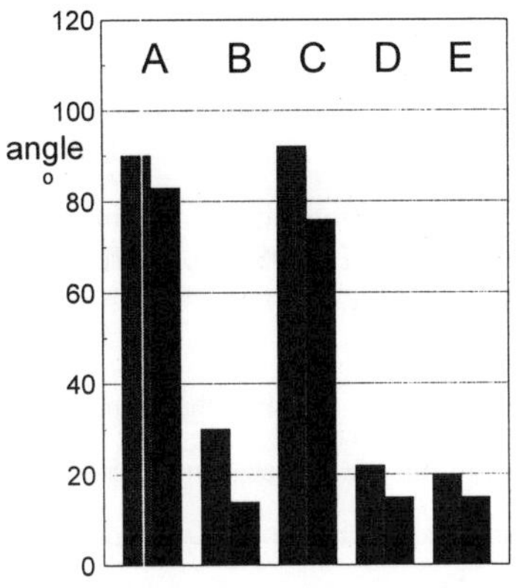

Fig. 6 Water advancing (left bars) and receding (right bars) contact angles for coatings deposited from plasma having following compositions (% of HDMSO, CF_4 and O_2, resp.): A 0,0,0; B 13, 87,0; C 50. 0, 50; D 14, 72, 14; E 7, 84, 7.

Figure 5 shows that, at least in the adopted experimental conditions (HMDSO/O_2/CF_4 4/50/4), the nature of the polymeric substrate has minor effect on the coating composition and thickness. Some preliminary contact angle measurements are reported in Fig. 6. Coatings obtained from HMDSO/CF_4 mixtures show low wettability, with contact angles similar to those of plain siloxane coatings (Ref. 13), while coatings from binary and ternary gas feed mixtures containing oxygen are all rather wettable.

Effects of discharge power on the deposition rate and surface composition of coatings are reported in Figs. 7 and 8, respectively. In these cases, the substrate polymer was HDPE. Figure 7 shows that the deposition rate depends directly on power discharge. Furthermore, it is also deeply affected by the nature of the polymerizable monomer. In fact, the deposition rate of HMDSO is always lower than that of TMDSO, the former being deposited only with a power

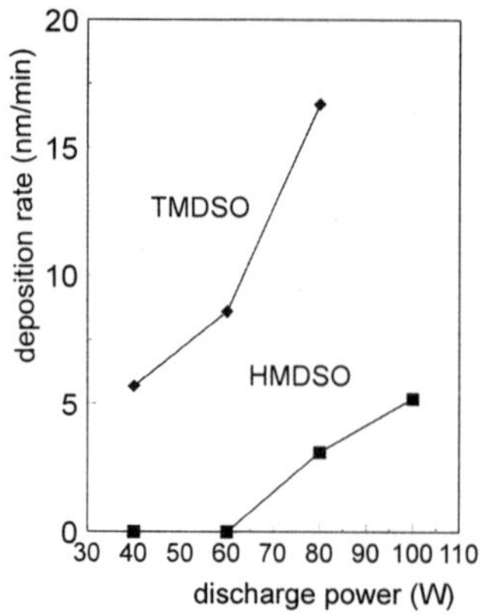

Fig. 7 Deposition rate of coatings from TMDSO and HMDSO on HDPE vs. discharge power.

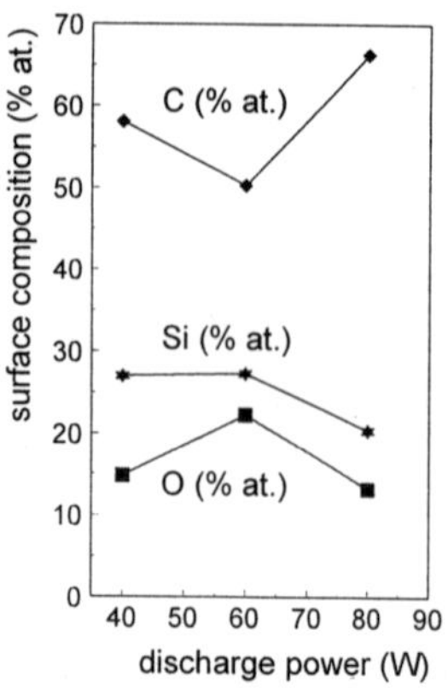

Fig. 8 Surface composition of coatings from TMDSO and HMDSO on HDPE vs. discharge power.

higher than 60 W. On the other hand, TMDSO produces, at high power (100 W), white flakes of some millimeters in diameter (Ref. 8). It is likely that such difference in reactivity depends on the fact that only in the TMDSO molecule hydrogen atoms are directly bonded to silicon atoms. Hydrogen atoms are mobile and thus easily abstracted from the disiloxane molecule, forming a reactive diradical (Ref. 14).

Surface composition of coatings obtained from TMDSO plasma polymerization at different discharge power values are reported in Fig. 8. At 40 W power, the experimental composition is very close to the theoretical (O 14.3% at., C 57.1% at., Si 28.6% at.). This result means that, in the adopted conditions, the TMDSO molecule does not undergo strong fragmentation (apart from hydrogen abstraction). At 60 W power, O/Si increases and C/Si decreases, in agreement with the expected increase in the inorganic character of the coating (Ref. 15). Finally, at 80 W, the opposite effect is observed, with an unexpected high carbon concentration. This result is interpreted considering that at high power (and, consequently, at high deposition rates), an inhomogeneous coating is deposited, with cracks through which the substrate remains exposed to the XPS analysis (Ref. 8).

The results of permeability measurements of HDPE coated by plasma polymerization of TMDSO are reported in Fig. 9. As expected, no effect on barrier properties against oxygen was observed, due to the organic character of the deposited layer. On the contrary, if the coating is subjected to post-treatment with O_2 plasma, capable of increasing its inorganic character, an

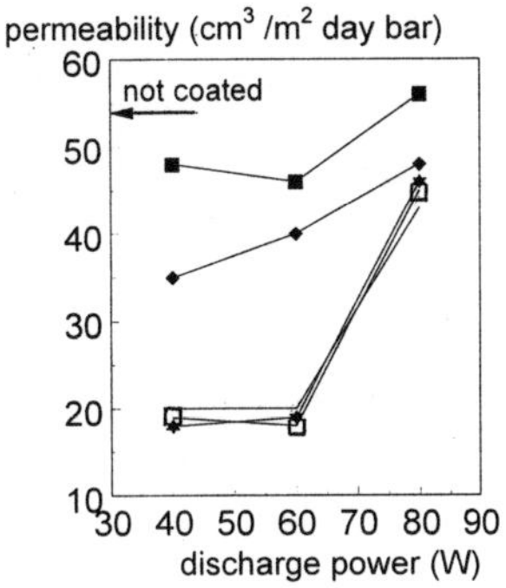

Fig. 9
Permeability to oxygen of TMDSO plasma-coated HDPE vs. discharge power after various post-treatment times with oxygen plasma.
(■ 0, ◆ 30, ∗ 60, □ 120, ◇ 600 min)

appreciable decrease in permeability was observed, at least for intermediate treatment power values (40-60 W), where the coating is sufficiently homogeneous (Ref. 8). Such a decrease is of the same order as that observed if oxygen was directly present in the gas feed (Ref. 16). The final result, however, is probably limited by the formation of cracks in the coating, due to the shrinkage of the molecular backbone (Ref. 8).

For coatings obtained with HMDSO/O_2/CF_4 mixtures, preliminary data show barrier properties similar to or better than those obtained with HMDSO/O_2. Also an improvement in wear resistance was observed, interesting for anti-scratch applications.

CONCLUSIONS

It was found that the inorganic/organic character of the deposited siloxane coatings has a great influence on the barrier properties. The occurrence of a SiO_2-like film free from defects and pinholes is essential for obtaining a good performance. This can be obtained both by adding oxygen in the plasma discharge or post-treating the deposited layer with oxygen plasma.

On the other hand, the use of ternary gas mixtures (siloxane/O_2/CF_4) allows to obtain deposited films with properties interesting both from the fundamental research and application points of view.

REFERENCES

(1) H. Yasuda, *Plasma Polymerization,* Academic Press, Orlando 1985.

(2) D. Korzec, D. Theirich, F. Werner, K. Traub, J. Engemann, *Surf. Coat. Technol.* **74-75**, 67 (1995).

(3) F. Garbassi, M. Morra, E. Occhiello, *Polymer Surfaces. From Physics to Technology,* Wiley, Chichester 1993.

(4) H.-C. Langowski, *Proc. Conf. "Metallocenes Europe '97"*, p. 233 (1997).

(5) L. Agres, Y. Ségui, R. Delsol, P. Raynaud, *J. Appl. Polym. Sci.* **61**, 2015 (1996).

(6) Y. Sawada, S. Ogawa, M. Kogoma, *J. Phys. D: Appl. Phys.* **28**, 1661 (1995).

(7) D.L. Cho, H. Yasuda, *J. Appl. Polym. Sci., Appl. Polym. Symp.* **42**, 73 (1988).

(8) M. Morra, E. Occhiello, F. Garbassi, *J. Appl. Polym. Sci.* **48**, 1331 (1993).

(9) J.A. Theil, J.G. Brace, R.W. Knoll, *J. Vac. Sci. Technol.* **A 12**, 1365 (1994).

(10) M.R. Alexander, R.D. Short, F.R. Jones, M. Stollenwerk, J. Zabold, W. Michaeli, *J. Mater. Sci.* **31**, 1879 (1996).

(11) D.L. Flamm, V.M. Donnelly, *Plasma Chem. Plasma Process.* **1**, 317 (1981).

(12) J.E. Coburn, *Plasma Chem. Plasma Process.* **2**, 1 (1982).

(13) Occhiello, *Angew. Makromol. Chem.* **222**, 189 (1994).

(14) Supiot, F. Callebert, O. Dessaux, P. Goudmand, *Plasma Chem. Plasma Process.* **13**, 539 (1993).

(15) Park and N. Kim, *J. Appl. Polym. Sci., Appl. Polym. Symp.,* **46**, 91 (1990).

(16) Inagaki, S. Tasaka, M. Makino, *J. Appl. Polym. Sci.,* **64**, 1031 (1997).

Macromol. Symp. ***139,*** *115–123 (1999)*

The Surface Tension of Polymer Liquids

Gregory T. Dee*, Bryan B. Sauer

Dupont Experimental Station, P. O. Box 80356
Wilmington, DE 19880-0356, USA

Abstract: A brief review of the surface tension of polymer liquids is presented. A strong emphasis is placed on recent measurements of surface tensions of homologous liquid series up to high-molecular-weight polymers, and the thermodynamic liquid properties of these same homologous series obtained from sources such as pressure-volume-temperature (PVT) data. The accuracy and limitations of the thermodynamic information which are used as input to many of the theories applied to the surface properties of polymer molecules are discussed. By scaling the surface tension data using a true measure of the cohesive energy density of the liquid state, we can clearly observe the entropic contribution to the surface tension caused by the conformational restriction of a large molecule at the liquid-vapor interface. The scaling implies the existence of a corresponding states principle for both polymer liquids and for low-molecular-weight liquids. The ramifications of the existence of a corresponding states principle for the surface tension of polymer melts are discussed. One consequence of the corresponding states principle is that it allows us to use surface tension measurements to compute the cohesive energy density of polymer melts using PVT data.

Introduction

The surface tension of polymer molecules has been studied for many years (Ref.1). Polymer melts have high viscosity which leads to long relaxation times for many of the standard methods for the measurement of the surface tension, γ. Many polymers have high melting temperatures or glass transition temperatures. These high temperatures make measurement difficult and the stability of the polymer at these temperatures is often such that degradation of the polymer occurs before a measurement can be made. Finally, polymers are seldom pure. They are usually polydisperse and can have structural defects such as branching. Therefore, much of the experimental literature has been confined to measurements of surface tension for low-molecular-weight liquids.

Numerous theoretical methods have been applied to the study of polymers at interfaces. In many cases the theoretical tools used to model the behavior of small molecule systems have been extended to study polymer systems. The models are of

 CCC 1022-1360/99/$ 17.50+.50/0

two types. The first are those which seek to describe the global properties of the liquid using molecular potentials. This is a very difficult task and currently only semiquantitative success has been achieved for simple short-chain molecules such as the alkanes. Such methods yield details about the microscopic details of the molecular conformations of macromolecules near liquid-vapor or liquid-liquid interfaces. A second more common and older approach is to coarse the grain of the system so that many of the microscopic details are lost and then use macroscopic thermodynamic information to calibrate the model. The model is then used to predict other macroscopic properties such as the surface tension. Such models are numerous and are not limited to the prediction of macroscopic variables. The square gradient theory (Ref. 2), for example, can predict the liquid-vapor interface dimension which is of the order of 10 - 15 Å. The power of such models is that they allow us to explore the meso scale structure of the surface. A combination of these two approaches can yield much information about the behavior of polymer molecules at interfaces and also yield practical tools for the prediction of macroscopic thermodynamic properties as well.

How are polymer molecules different from small molecules at or near an interface? In what follows we will focus on the liquid-vapor interface. Polymers form random coil conformations in the bulk liquid state. The spherical spatial domain occupied by the molecule is characterized by the radius of gyration, R_g, of the molecule. R_g is a function of the molecular weight of the molecule and its chemical composition. This restriction in the number of configurations of the molecule at the surface, will lead to an entropic contribution to the surface tension (Ref. 3). Other aspects of polymers which one might expect to affect surface properties are end groups of the polymer, side groups, and changes in chemical structure along the molecule, i.e. block copolymer structure. These and other structural features of polymer molecules contribute in varying degrees to surface properties of polymer liquids. The challenge is then to quantify these various contributions. A problem which immediately presents itself is the fact that the largest contribution to the surface tension will come from the cohesive energy of the liquid and unless we are able to predict this contribution with sufficient accuracy, we cannot measure the efficacy of any theory which claims to compute the magnitude of other contributions to the surface tension.

Scaling

The surface tension of complex molecules at a liquid vapor interface can be written in the following manner.

$$\gamma\,(T) = \gamma_{isotropic} + \gamma'_{non\text{-}local} + \gamma'''_{anisotropic} + \ldots \qquad (1)$$

The first term in the above expression is the contribution to the surface tension from the local interactions where we assume that the system is isotropic and disordered up to the interface. This term is the dominant contribution and the only one in many weakly interacting small-molecule liquids. The second term is the non-local contribution to the surface tension which can result from the presence of large-chain molecules at an interface. The molecules are conformationally constrained and thus even though most of the molecules may occupy the bulk liquid domains away from the interface, these conformational effects contribute an entropic contribution to the surface tension. The magnitude of this effect was predicted by Hong and Noolandi (Ref. 3) to be of the order of 10% of the value of the surface tension. The third contribution comes from situations where molecules order at the interface due to the strong directional interactions between them. This increase in order at the interface also results in an entropic contribution to the surface tension. Water is an example of a molecule which orders at the liquid-vapor interface in order to try to satisfy its hydrogen bonds with neighboring molecules at the interface. In the case of water, this contribution can be very large. Other contributions implied in Eq. 1 may exist whose size will be determined by both molecular interactions and molecular structure.

In order to study these various contributions to the surface tension we need to be able to partition them in some fashion. One way to achieve this is to choose molecules where only some of the above contributions are operative. In the case of weakly interacting polymers, for example, only the first two terms in Eq. 1 are dominant. If we now scale Eq. 1 using the isotropic contribution, we obtain the following simple expression:

$$\gamma(T) \,/\, \gamma_{isotropic} = F(T/T^*) + f'(T/T^*) \qquad (2)$$

where the first term on the right-hand side is a universal function for all molecules which form random interfaces and have no non-local contributions to the surface tension and f'(T/T*) represents the contributions from non-local effects. T* is an appropriate scale factor for the temperature. Our ability to separate these contributions depends on the degree to which F(T) and f′(T) are truly universal functions. Another way of stating this is that we are assuming that a corresponding states principle exists for the surface tension of molecules which are not extended in space and which do not form strong directional intermolecular bonds.

One approach to studying the non-local contributions contained in the function f′(T) is to study an oligomeric series. The surface tensions of linear-chain molecules of different molecular weights have been measured that span molecular sizes from small compact molecules to long-chain polymer molecules (Ref. 4). Molecules with short chains will not suffer conformational penalties at the interface. As the molecular weight grows, we expect the entropic contribution from these molecules to turn on as a function of increasing molecular weight. Therefore, a possible experiment would be to examine Eq. 2 for such a molecular weight series. Clearly, the above analysis will only be successful if we can compute $\gamma_{\text{isotropic}}$ accurately from the bulk thermodynamic properties of the liquid. It is to this question that we now turn our attention.

Polymer Thermodynamic Properties

Pressure, Volume, and Temperature Data

Many of the models use thermodynamic properties to calibrate their microscopic features. The energy and length scales of the model are fixed by using bulk equilibrium thermodynamic physical properties. This is frequently accomplished in the case of polymer liquids by using equations of state, i.e., a function which relates the density to the pressure and the temperature. The equation of state derived from a model is fit to pressure, volume, and temperature (PVT) data for a given sample and the fitting parameters (P*,T*, V*,..), related to the microscopic model parameters, are obtained which characterize the thermodynamic properties. The microscopic parameters of the model can be obtained by fitting any thermodynamic property derivable from the model to the experimentally determined thermodynamic property. If the model is a good fitting function for that thermodynamic property or properties, then the model should describe those properties with which it has been calibrated but it may not be able to predict other thermodynamic properties. This latter case is more frequently the case. Therefore, depending on the thermodynamic property one wishes to investigate using a model, one should play close attention to the information used to calibrate the model.

PVT data for polymer melts have been lacking in the past. Like surface tension measurements, the measurement of the density of these viscoelastic fluids is not trivial and only recently have dilatometers been developed (Ref. 5) to provide data. An important aspect of any model is therefore the ability to acquire an accurate description of these thermodynamic data. Most three-parameter equations of state cannot provide accurate descriptions of PVT data over large temperature and pressure domains (Ref. 6). One way to circumvent this difficulty is to allow the fitting parameters to be functions of temperature and pressure. In this way one can take full advantage of the thermodynamic information. The only purpose of the temperature-dependent parameters is to capture an accurate description of the thermodynamic properties of the sample. The parameters are not used to fit the surface tension data.

The Cohesive Energy Density (CED)

The cohesive energy density (CED) is defined as the ratio of the internal energy, U, to the volume, V (CED = U/V). It is a measure of the strength of the interactions between the molecules. The solubility parameter is defined to be the square root of the cohesive energy density. It plays a major role in the modeling of the thermodynamic properties of polymer blends and solutions. Hildebrand and Scott (Ref. 7) related the CED to the surface enthalpy and they observed a strong correlation between the surface tension and the CED for a number of small-molecule liquids. In the case of small-molecule liquids, it is possible to calculate the CED using the heat of vaporization. In the case of polymer liquids, which have vanishingly small vapor pressures, it is not possible to compute the CED in this fashion. Numerous other thermodynamic properties depend on the CED and hence a number of correlations exist which relate these properties to the CED.

The Internal Pressure

It was Hildebrand and Carter (Ref. 8) who pointed out the correlation between the CED and the internal pressure. Although not usually stated, many equations of state are formulated based on the idea that the internal pressure is equal to the CED. The quantity of physical interest to us is the cohesive energy density (CED) of the polymer melt under study. This quantity cannot be directly computed from PVT data. A quantity which can be computed directly from PVT data is the internal pressure (P_i) which is defined as follows:

$$P_i = (\partial U/\partial V)_T = T(\alpha/\beta) - P \quad (3)$$

where U is the internal energy, V is the volume, T is the absolute temperature, α is the volume expansivity, β is the isothermal bulk compressibility, and P is the pressure. Hildebrand and Carter (Ref. 8) noted that at room temperature, the ratio of the internal pressure to the CED was close to unity for numerous hydrocarbon liquids with varying degrees of complex chemical structure. For alkanes for example, this ratio, which Hildebrand denoted as n, (where, $n = P_i/CED$), varied between 1.05 and 1.12 in the range C5-C14. This implied that if one used P_i as an estimate of the CED for alkanes, one would overestimate the CED by 5 - 10 %. Clearly, if one knows the value of n, then one can calculate accurately the CED from the internal pressure. Therefore, P_i provides us with information on the cohesive forces of the melt with accuracy in the range of 5-15% for alkanes.

Figure 1 shows a plot of the ratio $n(T,M_n)$ ($n = P_i/CED$) for alkanes where P_i was determined using PVT data and the CED was determined from heat of vaporization data. The plot shows clearly that for molecules with more than 12 carbon atoms, n is independent of molecular weight. Therefore for high molecular weights, we can use this function, n(T), to determine the CED using the internal pressure of the liquid for that material.

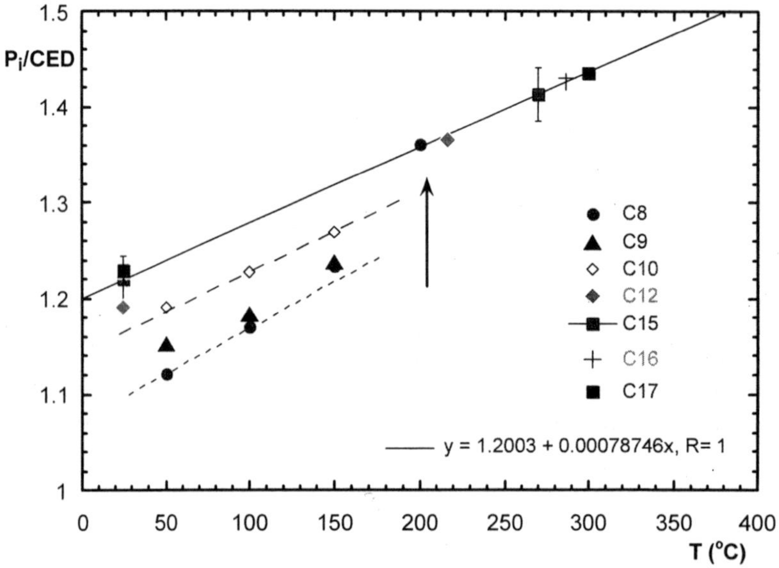

Fig. 1 Alkane thermodynamic data. A plot of $n(T,M_n)$ versus temperature for the alkanes

The Corresponding States Principle

Patterson and Rastogi (Ref. 9) tested the corresponding states principle for the surface tension of polymer melts. They scaled the surface tension using the bulk compressibility, β and volume expansivity, α. They defined a reduced surface tension $\gamma = \gamma/\gamma^*$, where $\gamma^* = \beta^{2/3}\,\alpha^{1/3}\,k^{1/3}$ and k is the Boltzmann constant. They found that plotting this reduced surface tension versus a reduced temperature, αT, the data for a number of polymers and small-molecule liquids fell on a universal curve with a scatter of 7-10%. The verification of the existence of a corresponding states principle depends on the availability of accurate bulk thermodynamic data for polymer and oligomer melts. Therefore, some of the scatter may be due to inaccuracies in the limited thermodynamic data.

By scaling in this manner, Patterson and Rastogi are using the internal pressure as an approximation to the CED of the liquid. This same information can be cast in many equivalent forms. For example, we could fit an equation of state to the PVT data for the liquid and use the reduction parameters of the model (P*,V*, and T*) as scaling variables where $\gamma^* = \alpha\, P^{*2/3}\, T^{*2/3}$ and T scaled with T*. The universal functions obtained will have different forms but the information content will be the same in each case (Ref 10). This work of ours showed that scaling the surface tension data for polymers and oligomers did not show any evidence of an entropic contribution to the surface tension. It also showed that using the internal pressure as a measure of the CED of the liquid state, we could predict the surface tension of numerous polymers with varying chemical structure to within 7 to 10 %. For molecules with the same chemical structure but with different molecular weights, we observed a strong corresponding states principle and could predict the surface tension to within 1-2 % using the internal pressure. This leads us to the conclusion that if we could scale the data with the CED for the liquid state, we could possibly collapse all the data for polymer liquids onto a single universal curve.

Given that we know the relation between the CED and the internal pressure for polyethylenes above C12, we can fit the equation of state directly to the CED and the density of the liquid. In doing so we are demanding that the equation of state give the correct CED and density at any given temperature and it is these properties of the bulk liquid that are most closely related to the surface tension. Figure 2 shows the surface tension of a number of alkanes scaled in this fashion. At low molecular weights, the data seem to fall on a lower curve and as the chain length increases, the data shift to a second higher curve with lower slope. This is the expected behavior predicted from our earlier simple picture of a polymer molecule at the liquid-vapor interface. Therefore, we expect very long chain molecules to lie on the upper curve and small molecules whose spatial dimensions are comparable to or smaller than the the interface width to scale on the lower curve. The magnitude of the effect is of the order of 10% of the surface tension as predicted in Ref. 3. These results demonstrate clearly that the simple idea of decoupling the contributions to the surface tension is possible in the case of polymer molecules. This fact opens up some new tools for the analysis of the thermodynamic properties of polymer melts which are enumerated in the conclusions.

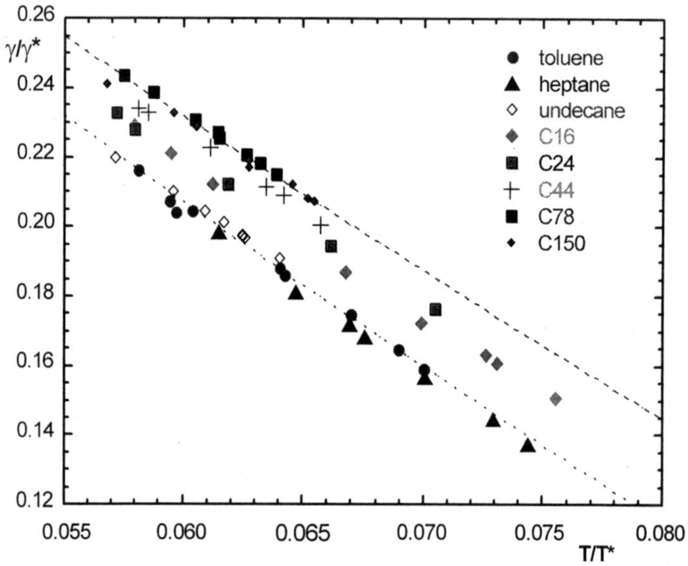

Figure 2. A plot of the scaled surface tension versus scaled temperature

Conclusions

The universal scaled surface tension provides a method for the calculation of the CED for a given polymer knowing its surface tension and the internal pressure of the melt determined from PVT data. By using the internal pressure to scale the surface tension data, we will obtain a scaled curve which will not correspond to the universal function we obtained for PE. If we assume that the curve we obtained for PE will be the same for all weakly interacting polymers, then the mapping of the curve scaled with the internal pressure onto the curve scaled with the CED specifies the relationship between the CED and the internal pressure. Using this approach, we will obtain different functions, n(T), for polymers with different chemical structure. This is a very powerful result since there is no direct method of measuring the CED for many polymer melts. We are currently using this method to obtain accurate thermodynamic properties of polymer melts.

We have demonstrated that one can decouple the entropic contribution to the surface tension of a polymer molecule from that which results from the local interactions between disordered molecules. Using accurate thermodynamic bulk properties and the corresponding states theory for surface tension, we can decouple the surface tension into those terms which satisfy the principle and those which do not. For chain molecules, two universal curves exist. One for low molecular weights and one for high. The difference between the curves is a manifestation of the non-local entropic contribution to the surface tension from polymer molecules at a liquid-vapor interface.

Acknowledgments
The authors would like to thank Ben Haley for assistance in measuring the PVT data and Nick DiPaolo for assistance in measuring the surface tension data.

References

(1) Wu, S., "Polymer Interface and Adhesion", Marcel Dekker, New York 1982.

(2) Poser, C.I.; Sanchez, I.C., *J. Colloid Interface Sci.* *69,* 539 (1979).

(3) Hong, K.M.; Noolandi, J., *Macromolecules 14,* 1223 (1981). Hong, K.M.; Noolandi, J., *Macromolecules* *14*, 1229 (1981).

(4) Dee, G.T.; Sauer, B.B., *J. Colloid Interface Sci. 152,* 85 (1992).

(5) Zoller, P. in: Mark, H.; Bikales, N.; Overberger, C.; Menges, G., Eds., "*Encyclopedia of Polymer Science and Engineering* ", 2nd ed., Vol. 5, p. 69, Wiley, New York 1986.

(6) Dee, G.T.; Walsh, D.J., *Macromolecules* *21*, 811(1988).

(7) Hildebrand, J.H.; Scott, R.L., "*The Solubility of Nonelectrolytes* ", 3rd ed., Reinhold Publishing Corp., New York 1950.

(8) Hildebrand, J.H.; Carter, J.M., *J. Am. Chem. Soc.* 54, 3592 (1932).

(9) Patterson, D.; Rastogi, A.K., *J. Phys. Chem.* *74*, 1067 (1970).

(10) Dee, G.T.; Sauer, B.B., *Polymer* 36, 1673 (1995).

Modification of the Polycarbonate/Poly(vinylidene Fluoride) Interface by Poly(methyl Methacrylate). Effect on the Interfacial Adhesion and Interfacial Tension

Noureddin Moussaif, Robert Jérôme*

Center for Education and Research on Macromolecules (CERM), University of Liège, Institute of Chemistry, B6, Sart-Tilman, 4000 Liège, Belgium

Abstract: Polycarbonate (PC) and poly(vinylidene fluoride) (PVDF) are two immiscible polymers which form two-phase blends with weak interfacial adhesion and high interfacial tension. This situation may be changed by the addition of poly(methyl methacrylate) (PMMA), which concentrates preferably in the PVDF-rich phase, but also at the PVDF/PC interface. The interfacial activity of PMMA was estimated by the measurement of the interfacial adhesion and interfacial tension in relation to the PMMA content in the PVDF/PC blends. The interfacial adhesion between PC and homogeneous PVDF/PMMA blends of various compositions was measured by the dual cantilever beam technique. The imbedded fiber retraction method was used for the measurement of the interfacial tension. A very beneficial effect was observed when PVDF was premixed with PMMA amounts increasing up to ca. 35 wt.-%. Beyond that content, the improvement tends to level off.

INTRODUCTION

Addition of properly designed block or graft copolymers to immiscible polymer blends is a well-known strategy to improve the phase dispersion, to stabilize the phase morphology and to improve the interfacial adhesion (Refs. 1,2). However, the cost of preformed block or graft copolymers and the limitation of reactive processing in the *in situ* formation of block copolymers (Refs. 3-5) are incentives to consider new compatibilization strategies. This problem is still more acute in the case of polycarbonate (PC) and poly(vinylidene fluoride) (PVDF) blends since no parent block or graft copolymers can be made whatever the synthetic approach is. In this work, poly(methyl methacrylate) (PMMA) has been considered as a potential compatibilizer for PC/PVDF polyblends, since PMMA is known to be miscible with PVDF (Refs. 6-8) and compatible with PC (Refs. 9-11). In the extreme, PMMA might behave as a common "solvent" for PC and PVDF in the melt. Before testing the ability of PMMA to improve the PC/PVDF interfacial adhesion and to decrease the PC/PVDF interfacial tension, it is worth noting that PC has been toughened by core-shell rubber particles with a PMMA

 CCC 1022-1360/99/$ 17.50+.50/0

shell. The compatibility of this PMMA shell with PC was proposed to provide an interfacial adhesion strong enough for the rubber particles to cavitate rather than to be debonded (Ref. 12). Another valuable observation was reported by Hobbs et al. (Ref. 13), who showed by transmission electron microscopy that PC envelopes PMMA domains in a ternary PC/PMMA/PBT blend, in agreement with the spreading coefficients calculated from interfacial tension data.

For PMMA to have an interfacial activity, it has to be located at the interface of the two immiscible polymers. Although experimental techniques for the surface analysis, such as forward recoil spectroscopy, secondary ion mass spectroscopy and neutron reflectrometry, can be used, they require a preliminary separation of the phases and, for some of these techniques, a labeling of one polymeric component. It might thus be more straightforward and convenient to measure properties in a direct connection with the real interface, i.e., the interfacial tension by the imbedded fiber retraction (IFR) method (Refs. 14-16), and the interfacial adhesion by the dual cantilever beam test (Ref. 17). The purpose of this work is thus to investigate how the PC/PVDF interfacial tension and adhesion are improved by premixing PVDF with increasing amounts of PMMA.

EXPERIMENTAL

Materials

Table 1 lists the source, molecular weight, polydispersity and glass transition of the polymers used in this study.

Table 1 Main characteristics of the investigated polymers

Polymer	**Trade name**	**Source**	**M_w** $\times 10^{-3}$	**M_w/M_n**	**T_g**[a] °C
Polycarbonate (PC)	Makrolon 3103	Bayer	58 [b]	1.7 [b]	150
Poly(methyl methacrylate) (PMMA)	Diakon	ICI	60 [b]	1.6 [b]	118
Poly(vinylidene fluoride) (PVDF)	Solef x10N	Solvay	125	1.8	-45
Poly(vinylidene fluoride) (PVDF)	Solef 1008	Solvay	110	1.8	-45

[a] By dynamic mechanical analysis at 1 Hz; [b] by GPC in THF using polystyrene standards.

Measurement of the interfacial adhesion

The fracture toughness of the phase boundary, G_c, was measured using the dual cantilever beam method (Ref. 17). Two-mm thick plates of PC and PVDF/PMMA blends were superposed and compression-molded at a desired temperature. After a well defined period of time, the sandwich was cooled down under a small pressure and test samples of 0.9-1 cm width were cut out. A razor blade was inserted between the two constitutive layers, and the length of the crack propagating ahead the razor blade was measured with a travelling microscope after at least 84-h equilibration at room temperature. The toughness of the interface was calculated from Eq. 1 proposed by Creton *et al.* (Ref. 17) in the case where no component of the sandwiched assembly was previously glued on a rigid glass plate.

$$G_c = \frac{3d^2 E_1 h_1^3 E_2 h_2^3}{8l^4 \left(E_1 h_1^3 + E_2 h_2^3\right)} \quad [1]$$

where E and h are the modulus and thickness of the constitutive layers 1 and 2, d is the thickness of the razor blade, and l is the crack length. Since the moduli of the two plates were comparable, the use of a symmetric plate assembly allowed the crack to propagate right along the interface.

The Young modulus was calculated from the initial slope of the stress - strain curves. For this purpose, PVDF/PMMA blends were prepared by melt mixing with a two-roll mill (SCAMIA), at 200 °C for 10 min. 1 mm thick samples of PC and PVDF/PMMA blends were prepared by compression molding at 220 and 200 °C, respectively. Stress - strain curves for the PVDF/PMMA blends were recorded with an Instron universal tensile tester (model DY24) at a tensile rate of 20 mm/min.

Measurement of the interfacial tension

The interfacial tension between melted PC and PVDF/PMMA blends was measured by the IFR technique (Refs. 14-16), based on the measurement of the microscopic tracking of short imbedded fibers. The interfacial tension was calculated from the time dependence of the retraction process, as detailed elsewhere (Refs. 15,16). Briefly, the length and the effective radius (R) of the fiber were measured as functions of time, and the function $f(R/R_0)-f(R_e/R_0)$ was plotted vs. the retraction time t, in agreement with Eq. 2.

$$f(R/R_0) - f(R_e/R_0) = t\gamma_{12}/[((\eta_m+1.7\eta_f)/2.7)R_0] \quad [2]$$

where γ_{12} is the interfacial tension, R and R_e are the effective radii of the retracting fiber at time t and t=0, respectively. R_0 is the radius of a sphere of the same volume as the fiber, and η_m and η_f are the zero-shear viscosities of the matrix and the fiber, respectively. Equation 3 is the expression of functions $f(R/R_0)$ and of $f(R_e/R_0)$, whose values are available in ref. 15.

$$f(x) = \frac{3}{2}\ln\left[\frac{\sqrt{1+x+x^2}}{1-x}\right] + \frac{3^{3/2}}{2}\tan^{-1}\left[\sqrt{3}\frac{x}{2+x}\right] - \frac{x}{2} - \frac{4}{x^2} \quad [3]$$

where $x = R/R_0$ or R_e/R_0.

PC fibers were prepared by extrusion at 240 °C through the die of an equipment used for the measurement of melt indices. The polymer was previously dried at 120 °C for 24 h under vacuum. The final diameter was 0.1 mm.

The PMMA/PVDF matrices were prepared by melt mixing of PMMA and PVDF with a two-roll mill (SCAMIA) at 200 °C for 5 min. 0.5-mm plates of PMMA, PVDF and PMMA/PVDF blends were prepared by compression molding at 200 °C. The fiber and matrix were annealed at 150 and 100 °C, respectively, under vacuum overnight.

Short-section fibers (ca. 0.5 mm) were sandwiched between two plates of the same PVDF/PMMA blend, and this sandwich was placed between preheated glass slides. The samples were then transferred to an Orthoplan microscope equipped with a Mettler FP-5 hot stage and annealed at 150 °C for 30 min in order to allow PMMA to migrate to the interface. The fiber retraction was then measured at 220 °C.

The melt viscosity vs. frequency curves for all the PC, PVDF, PMMA and PVDF/PMMA blends were measured using a Bohlin constant-stress rheometer at 220 °C. Measurements were performed using the plate - plate configuration and a gap of approximately 1 mm, in the frequency range of 10^{-2}-10^{2} Hz.

RESULTS AND DISCUSSION

Figures 1 and 2 show the frequency dependence of the melt viscosity of PC, PVDF, PMMA and PVDF/PMMA blends of various compositions at 220 °C. The zero-shear viscosity (η_0) has been calculated from Eq. 4 based on the Yasuda model (Ref. 18).

$$\eta = \eta_0\left[1+\left(\alpha\dot{\gamma}\right)^a\right]^{(n-1)/a} \quad [4]$$

where α is a molecular constant (Ref. 19) equal to $12\eta_0/\pi^2\nu kT$, ν is the number of molecules per unit volume, kT is the usual product of the absolute temperature and the Boltzmann constant, n and a are the constant parameters of this model, which have to be optimized for fitting the experiment.

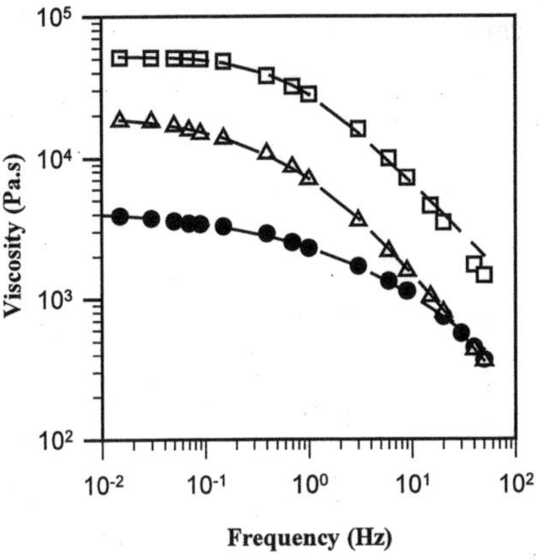

Fig. 1 Viscosity/frequency curves for PC, PVDF, and PMMA at 220 °C. The experimental data are fitted by the Yasuda model (broken line). □ PC, ● PVDF, △ PMMA.

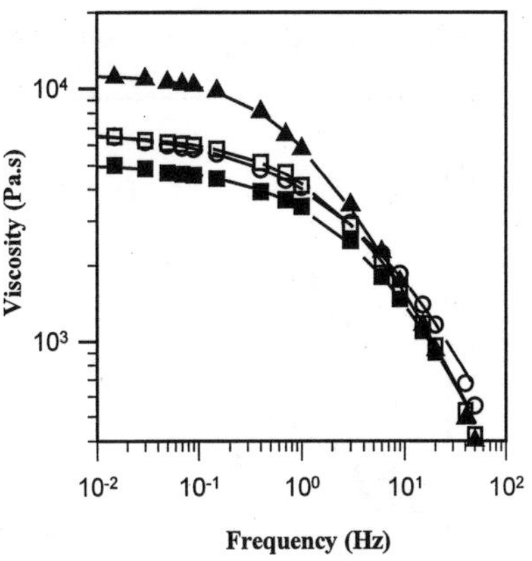

Fig. 2. Viscosity/frequency curves for PVDF/PMMA blends at 220 °C. The experimental data are fitted by the Yasuda model (broken line). ○ 80/20, ■ 60/40, □ 40/60 and ▲ 20/80.

Since the complex viscosity $\eta^*(f)$ should be numerically equal to the viscosity $\eta(\dot{\gamma})$, when the shear rate $\dot{\gamma}$ and frequency f are the same (Cox-Merz rule) (Ref. 20), the values of $\dot{\gamma}$ used in Eq. 4 are numerically equal to the experimental frequency. The values of η_0 are reported in Table 2 and used in Eq. 2 for the calculation of the interfacial tension.

Table 2 Zero-shear viscosity (η_0) of homopolymers and PMMA/PVDF blends

Material	η_0 at 220 °C (kPa.s)
PC	51.3
PVDF	4.4
PMMA	20.4
20 PMMA / 80 PVDF	6.9
40 PMMA / 60 PVDF	5.1
60 PMMA / 40 PVDF	6.6
80 PMMA / 20 PVDF	11.3

Plots of the function $F = f(R/R_0) - f(R_e/R_0)$ vs. the retraction time are shown in Fig. 3 for the systems under investigation. The expected linearity is actually observed, with a slope proportional to the interfacial tension. The experimental values of γ_{12} are listed in Table 3.

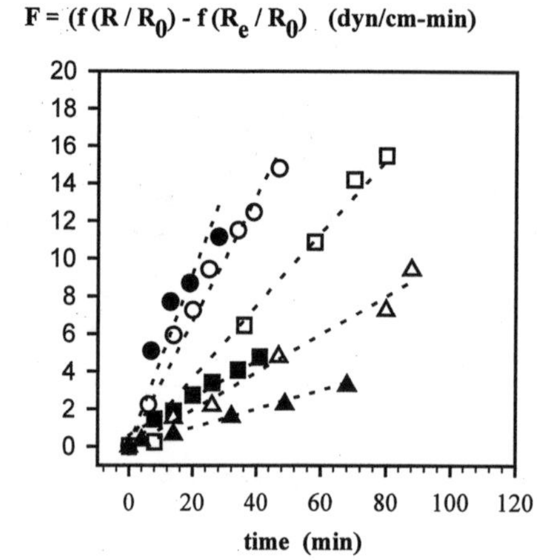

Fig. 3 Typical plot of the function $F=[(f(R/R_0) - f(R_e/R_0)]$ vs. the retraction time for a PC fiber in PVDF, PMMA and PVDF/PMMA blends:
● 100/0, F= 0.46 t; ○ 80/20, F= 0.333 t; ■ 60/40, F= 0.099 t; □ 40/60, F = 0.193 t; △ 20/80, F = 0.101 t; ▲ 0/100, F = 0.05 t.

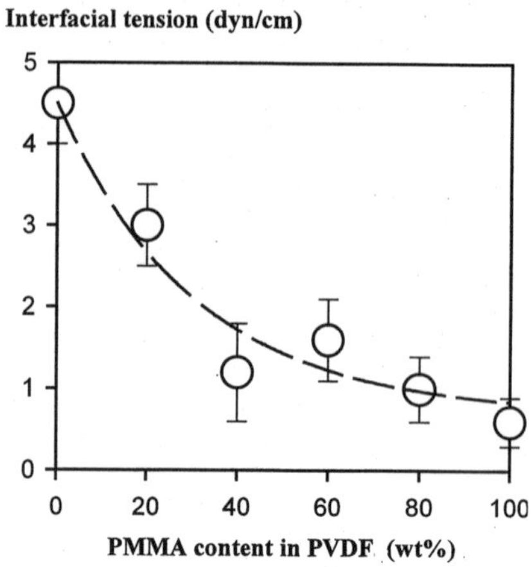

Fig. 4 Plot of the interfacial tension between PC and PVDF/PMMA blends as a function of the weight percentage of PMMA in PVDF (T=220 °C).
○ experimental data, – – the fitting curve from Eq. 6.

Table 3 Interfacial tension between PC and PMMA/PVDF blends of different compositions at 220 °C

Matrix	γ_{12} (dyn/cm)
PVDF	4.5 ± 0.6
20 PMMA / 80 PVDF	3.0 ± 0.5
40 PMMA / 60 PVDF	1.2 ± 0.5
60 PMMA / 40 PVDF	1.6 ± 0.5
80 PMMA / 20 PVDF	1.0 ± 0.4
PMMA	0.6 ± 0.3

The PC/PVDF interfacial tension (4.5 ± 0.6 dyn/cm) is much higher than the PC/PMMA interfacial tension (0.6 ± 0.3 dyn/cm), consistently with the complete immiscibility of PC and PVDF and the well-known PC-PMMA compatibility. These experimental data are also in a pretty good agreement with the values predicted by the harmonic mean equation (Ref. 21) on the basis of the surface tension of each constituent polymer (Eq. 5).

$$\gamma_{12} = \gamma_1 + \gamma_2 - \frac{4\gamma_1^d \cdot \gamma_2^d}{\gamma_1^d + \gamma_2^d} - \frac{4\gamma_1^p \cdot \gamma_2^p}{\gamma_1^p + \gamma_2^p} \qquad [5]$$

where the d and p superscripts refer to the dispersive and polar components of the surface tension, respectively. From the values reported in the scientific literature for the surface tension of PC, PMMA and PVDF (Ref. 22,23), $\gamma_{PC/PVDF} = 3.8$ dyn/cm and $\gamma_{PC/PMMA} = 0.5$ dyn/cm at 220 °C have been calculated for the PC/PVDF and PC/PMMA pairs, respectively.

Figure 4 clearly shows that the high interfacial tension of the immiscible PC/PVDF pair at 220 °C decreases rapidly on premixing PVDF with increasing amounts of PMMA. Beyond ca. 40 wt.-% PMMA in PVDF, this effect tends to level off.

Anastasiadis et al. (Ref. 24) have studied the compatibilizing effect of a diblock polystyrene-*block*-1,2-polybutadiene in polystyrene/1,2-polybutadiene blends. They have observed that the interfacial tension was decreasing as the concentration of the diblock copolymer was increased according to a relationship quite similar to that one reported in Fig. 4. On the basis of the experimental data for the polystyrene/1,2-polybutadiene blends, Tang and Huang (Ref. 25) have proposed a theoretical equation accounting for the dependence of the interfacial tension (γ) on the content of the compatibilizer (Eq. 6).

$$\gamma = (\gamma_0 - \gamma_s) \cdot e^{-KC} + \gamma_s \qquad [6]$$

where K is the rate constant for the change in the interfacial tension with the concentration of the compatibilizer (C) and γ_0, γ, γ_S are the interfacial tension in the absence of the compatibilizer, in its presence, and for the completely saturated interface, respectively.

Equation 6 fits very well the experimental data reported in this study for the PC/PVDF system, where a homopolymer (PMMA) is responsible for the compatibilization effect rather than a traditional diblock (or graft) copolymer (Fig. 4). Equation 7 is the expression of Eq. 6 in the particular case of the PC/PVDF two-phase system modified by PMMA.

$$\gamma = 3.8\,e^{-0.0328c} + 0.7 \qquad [7]$$

It thus appears that Eq. 6 might be of a general applicability.

From Fig. 4, it might be qualitatively concluded that as soon as premixed with PVDF, PMMA basically migrates to the interface and decreases very efficiently the interfacial tension. When enough PMMA is premixed with PVDF to saturate the interface, hence for substituting a PMMA/PC interface for a PVDF/PC interface, the effect tends to level off. In an essentially empirical way, the tangents at each origin of the curve fitting Eq. 6 intersect at a PMMA content in the PVDF/PMMA blend of ca. 35 wt.-%. This value might be considered as the concentration of the compatibilizer required for the saturation of the interface.

The interfacial adhesion between PC and PVDF containing increasing amounts of PMMA has also been measured under standard conditions (Ref. 26) (time = 10 min, T = 200 and 220 °C). Figure 5 shows that as long as the PMMA content in PVDF does not exceed 15 wt.-%, the fracture toughness is too small to be measured accurately. Beyond that PMMA content, the interfacial adhesion increases rapidly reaching a plateau value, which is nothing but the interfacial adhesion measured between PC and neat PMMA. Once again the intersection of the tangents at the two extremities of the experimental curve defines an initial PMMA content in PVDF of ca. 35.-wt % in a close agreement with the value extracted from the interfacial tension data.

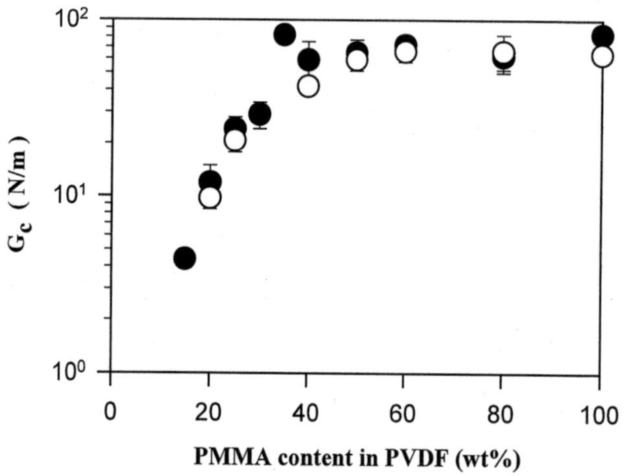

Fig. 5 Dependence of the fracture toughness (G_C) on the PMMA content in PVDF for the PC/PVDF assembly (● T = 220 °C, o T = 200 °C).

The experimental dependence of both the interfacial tension and the interfacial adhesion on the PMMA content in the PVDF/PMMA blends clearly shows that PMMA migrates and accumulates at the PC/PVDF interface more rapidly than its nominal concentration in PVDF would suggest. Indeed, a PC/PMMA interface is completely substituted for the original PC/PVDF interface when ca. 35 wt.-% PMMA is mixed with PVDF. The thermodynamic driving force for the preferential localization of PMMA at the interface has to be found in the smaller PC/PMMA interfacial tension (0.6 dyn/cm, Table 3) compared with the PC/PVDF value (4.5 dyn/cm, Table 3).

Although some intermolecular reactions might be suspected to occur between PC and PMMA at 220 °C (Ref. 27), the experimental data in Fig. 5 remain unchanged when the samples are molded and tested at 200 °C. This observation indicates that the measurements carried out at 220 °C are quite reliable. Furthermore, the plateau value for the interfacial adhesion between PC and the 50/50 PMMA/PVDF blend remains quite constant over long periods of time (16 h) at 200 °C, which agrees with a good thermal stability and the absence of PMMA and PC interdiffusion at this temperature.

In order to confirm that the substitution of a PMMA/PC interface for the original PVDF/PC interface is responsible for the improved interfacial adhesion, this property has been measured at 220 °C for a trilayer PC/PMMA/PVDF assembly prepared by spinning preliminarily PMMA onto PVDF and molding the final assembly for 10 min at 220 °C. In this case, the fracture toughness is 9 ± 1 N/m, hence as low as for the system prepared by premixing 20 wt.-% PMMA in PVDF (Fig. 5). This observation proves that the trilayered PC/PMMA/PVDF system is so far from a thermodynamic equilibrium that PMMA rapidly diffuses into PVDF and leaves the PVDF/PC interface only partly covered. This diffusion is no longer observed when the 60/40 (wt/wt) PVDF/PMMA is similarly spin-coated with PMMA before assembling with PC. After molding at 220 °C for 10 min, the interfacial toughness (98 ± 6 N/m) is very close to the value reported in Fig. 5 for the same system without the intermediate layer of PMMA. Since the spin-coated PMMA layer maintains the interfacial adhesion, it has to interconnect PVDF and PC more likely as a result of interdiffusion in the PVDF/PMMA phase. At this stage of the research, no conclusion can be drawn about the concentration of PMMA chains in the PC matrix, although this would be the expected way for creating good adhesion (Refs. 28-30). It is, however, worth pointing out that Yoon and Han (Ref. 31) have measured a PC/PMMA interfacial thickness of 32.5 and 29.6 Å at 180 °C and 200 °C, respectively, by small-angle neutron scattering for a ca. 1000-min annealing time.

CONCLUSIONS

The interfacial tensions between PC and PVDF as well as PMMA and a series of PVDF/PMMA blends have been measured by the imbedded fiber retraction method. The high interfacial tension between PC and PVDF rapidly decreased by premixing ca. 35 wt.-% PMMA with PVDF. In parallel, the PC/PVDF interfacial adhesion increased as rapidly as the interfacial tension decreased in relation to the PMMA content in the PVDF/PMMA blends. These observations are consistent with the interfacial activity of PMMA at the PVDF/PC interphase, as confirmed by the validity of the model proposed by Tang and Huang, accounting for the interfacial activity of a diblock copolymer in two-phase polymer blends. The best reinforcing effects appear to take place when the PMMA/PC interface is completely substituted for the original PVDF/PC interface.

ACKNOWLEDGMENT

The authors are very grateful to the *Services Fédéraux des Affaires Scientifiques, Techniques et Culturelles* in the framework of the "Pôles d'Attraction Interuniversitaires: Chimie et Catalyse Supramoléculaire". N. M. is indebted to the *Ministère de l'Enseignement Superieur, de la Formation des Cadres et de la Recherche Scientifique du Maroc* for a fellowship. He wishes to thank Mr. C. Pagnoulle and Dr. P. Maréchal for fruitful discussions, and Mr. R. Pirard for help in using the Bohlin rheometer.

REFERENCES

1. R. Fayt, R. Jérôme, Ph. Teyssié, *Makromol. Chem.* **187**, 837 (1986).
2. D. R. Paul, S. Newman, Eds., in *Polymer Blends,* Vol. 2, Chap. 12 (1978).
3. N. C. Liu, W. E. Baker, *Adv. Polym. Technol.* **11**, 249 (1992).
4. M. Xanthos, S. S. Dagli, *Polym. Eng. Sci.* **31**, 929 (1991).
5. D. Curdo, A. Valenza, F. P. Lamantia, *J. Appl. Polym. Sci.* **39**, 865 (1990).
6. H. Tomura, H. Saito, T. Inoue, *Macromolecules* **25**, 1611 (1992).
7. R. E. Prud'Homme, *Polym. Eng. Sci.* **22**, 90 (1982).
8. P. J. Flory, *J. Am. Chem. Soc.* **87**, 1833 (1965).
9. W. N. Kim, C. M. Burns, *Macromolecules* **20**, 1876 (1987).
10. J. S. Chiou, J. W. Barlow, D. R. Paul, *J. Polym.Sci., Polym. Phys. Ed.* **25**, 1459 (1987).
11. Z. G. Gardlund, *Adv. Chem. Ser.* **206**, 129 (1984) (Polymer Blend and Composites in Multiphase Systems)
12. C. Cheng, A. Hiltner, E. Baer, P. R. Soskey, S. G. Mylonakis, *J. Appl. Polym. Sci.* **52**, 177 (1994); **55,** 1691 (1995).
13. S. Y. Hobbs, M. E. J. Dekkers, V. H. Watkins, *Polymer* **29**, 1598 (1988).
14. J. Craig, C. J. Carriere and A. Cohen, *J. Rheol.* **35**, 205 (1991).
15. C. J. Carriere, A. Cohen, C. B. Arends, *J. Rheol.* **33**, 681 (1989).
16. Cohen, C. J. Carriere, *Rheol. Acta* **28**, 223 (1989).
17. Creton, E. J. Kramer, C. Y. Hui, H. R. Brown, *Macromolecules* **25**, 3075 (1992).
18. J. M. Dealy and K. F. Wissbrun, in *Melt Rheology and its Role in Plastics Processing. Theory and Applications*, Chap. 3, van Nostrand Reinhold, New York 1990.
19. F. Bueche, S. W. Harding, *J. Polym. Sci.* **32**, 177 (1958).
20. W. P. Cox, E. H. Merz, *J. Polym. Sci.* **28**, 619 (1958).
21. S. J. Wu, *J. Polym. Sci: Part C* **34**, 19 (1971).
22. S. Wu, in *Polymer Interface and Adhesion*, Marcel Dekker, New York 1982.

23. J. Brandrup, E. H. Immergut (Eds.), *Polymer Handbook*, 3rd ed., J. Wiley and Sons, New York 1989.

24. S. H. Anastasiadis, I. Gancarz and J. T. Koberstein, *Macromolecules* **22**, 1449 (1989).

25. T. Tang, B. Huang, *Polymer* **35**, 281 (1994).

26. N. Moussaif, Ph. Maréchal, R. Jérôme, *Macromolecules* **30**, 658 (1997).

27. Debier, J. Devaux, R. Legras, *J. Polym. Sci., Part A: Polym. Chem.* **33**, 407 (1995).

28. K. Jud, H. H. Kausch and J. G. Williams, *J. Mater. Sci.* **16**, 204 (1981).

29. H. R. Brown, *Annu. Rev. Mater. Sci.* **21**, 463 (1991).

30. H. R. Brown, *Annu. Rev. Mater. Sci.* **21**, 463 (1991).

31. H. Yoon, C. C. Han, *Polym.Eng.Sci*. **35**, 1476 (1995).